RECENT PROGRESS IN SILICON-BASED SPINTRONIC MATERIALS

Series on Spintronics

Series Editors: Ching-Yao Fong *(University of California, Davis, USA)*
Lin H. Yang *(Lawrence Livermore National Lab., USA)*

SERIES ON SPINTRONICS – VOL. 1

RECENT PROGRESS IN SILICON-BASED SPINTRONIC MATERIALS

L. Damewood & C. Y. Fong
UC Davis

L. H. Yang
Lawrence Livermore National Laboratory, USA

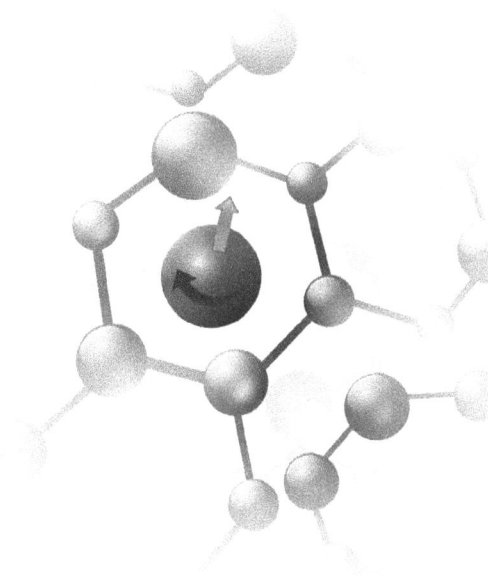

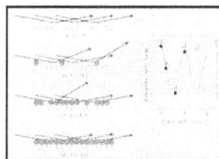

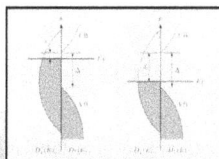

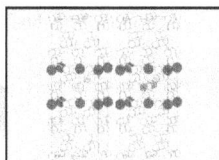

World Scientific

NEW JERSEY · LONDON · SINGAPORE · BEIJING · SHANGHAI · HONG KONG · TAIPEI · CHENNAI

Published by

World Scientific Publishing Co. Pte. Ltd.

5 Toh Tuck Link, Singapore 596224

USA office: 27 Warren Street, Suite 401-402, Hackensack, NJ 07601

UK office: 57 Shelton Street, Covent Garden, London WC2H 9HE

British Library Cataloguing-in-Publication Data

A catalogue record for this book is available from the British Library.

Series on Spintronics — Vol. 1
RECENT PROGRESS IN SILICON-BASED SPRINTRONIC MATERIALS

ISBN 978-981-4635-99-8

In-house Editor: Rhaimie Wahap

To our families

Preface

The idea of "spintronics" is to utilize the spin degree of freedom as the operational paradigm for the transport and storage of information in devices. An extremely important subset of spintronics is the application of ferromagnetic metals in devices such as magnetic disk read-heads and magnetic memory. Another branch of spintronics, put forward in the 1990s, aims to integrate spins into the semiconductor-based device industry. However, no device has been successfully made to operate at room temperature. Ideal materials for making spintronic devices have been identified as half-metals: a material with one of its spin channels behaving like a metal while the oppositely oriented spin channel exhibiting insulating properties. The unique feature to be used for spintronics is that the current in these materials are 100% spin polarized. Full- and half-Heusler alloys, oxides, such as CrO_2 and transition metal pnictides with simple zinc blende structure have been pursued, but due to difficulties during growth, structural transformations, and spin-flip transitions, these materials have not been shown to sustain half-metallic properties above room temperature.

Around 2005, several groups around the world started to explore spintronic materials in silicon (Si) because the most mature technologies developed for Si can be utilized to resolve some of the issues identified above. Since this is an unexplored research area, we realized that it will be more efficient to make use of the well-developed first-principles methods, based on density functional theory, to design new spintronic materials, then to select the ones with the most favorable properties for device applications to be grown. Even though there still needs to be an effective way to grow the materials—in particular layered samples—we have accumulated a significant amount of experience concerning first-principles methods, especially the development of physical pictures for the properties of and mechanisms resulting in half-metallicity. It is our desire to summarize our experience by focusing on the Si-based spintronic materials as examples to provide background for graduate students and industrial researchers working in this field so that they can easily join the efforts to reach the ultimate goal—the realization of spintronic devices.

These lecture notes begin with several foundations of condensed matter physics crucially relevant to the design and the issues related to the growth of Si-based spintronic materials. Describing the crystal structure is extremely important for understanding the magnetic properties of potential spintronic devices. The growth and characterization methods of crystals or multilayer devices require an in-depth understanding of the microscopic picture of crystals and their interactions with external fields. Furthermore, it is the mutual interaction of electrons through their spin and the Coulomb force of the exchange interaction that produces magnetic properties in crystals. We examine the exchange interaction—the principal interaction forming magnetic and half-metallic properties—and the spin–orbit interaction, which prevents half-metallicity in some materials. Developing Si-based spintronic materials requires physical insight to how controlling the strength of the interactions give rise to desirable properties, and which interactions, such as the spin–orbit interaction, deteriorate the quality of the half-metallic properties in materials.

In chapter 2, we present a number of first-principles methods, based on density functional theory, that are commonly used to facilitate the design of spintronic materials. The approximations characterizing the electron-electron exchange-correlation used in the first-principles methods are well-known to underestimate the fundamental gap of semiconductors and insulators, so we additionally present the so-called GW approximation. This method accounts for the different screening effects from the insulating and metallic channels of half-metals. It is extremely important that spintronic materials used in devices operate at room temperature. We thoroughly examine the methods of calculating the Curie temperature—above which ferromagnetic properties disappear—using quantities determined by first-principles calculations. For the experimental aspects, we present several methods and schemes for growing and characterizing half-metallic samples. The characterizing techniques are divided into three important aspects of device applications: the characterization of the crystal structure, the transport properties of electrons and the magnetization.

In chapter 3, we present the recent progress made developing Si-based spintronic materials, including our own designs. Physical pictures are emphasized to facilitate the understanding of the terminologies and physical mechanisms.

We are grateful for the support of the National Science Foundation with Grants ESC-0225007 and ECS-0725902. Work at Lawrence Livermore National Laboratory was performed under the auspices of the U.S. Department of Energy under Contract DE-AC52-07NA27344. Portions of this manuscript are derived from L. Damewood's Ph.D thesis (2013).

L Damewood and C Y Fong
University of California, Davis
L H Yang
Lawrence Livermore National Laboratory

Contents

Chapter 1

Spin-based Materials

1.1 Introduction

1.1.1 *History*

Integrated circuits (ICs) are devices consisting of many electronic circuits on a semiconducting (typically silicon) wafer, or chip. The number of transistors per unit area that can be integrated into a silicon (Si) chip is governed well by Moore's law. In 1965, Moore predicted an exponential trend in the number of transistors that could fit on a chip that would last at least until 1975. While this trend has continued much longer than originally predicted, there is a definite limit to packing components on a chip due to the finite interatomic distance of a few Angstrom (Å). For storage, similar trends exist for the number of magnetic domains on a magnetic hard drive. The capacity of hard drives has increased, into the terabytes, from a technological boost with the 1988 discovery by Baibich *et al.* (1988) of a magnetic layered structure, formed by iron (Fe) and chromium (Cr), exhibiting giant magnetoresistance (GMR). GMR-based devices, composed of layers of transition metal elements (TMEs) separated by a thin insulator, use the electron spin degree of freedom to manipulate, store and transmit information. In their experiments, Baibich *et al.* (1988) and Binasch *et al.* (1989) showed that layers of ferromagnetic (FM) and normal metal (NM) materials would show "giant" differences in the overall resistance depending on the magnetic alignment of the ferromagnetic layers.

Inspired by use of the spin degree of freedom in metallic device, Wolf introduced the term "spintronics", while at the U.S. Defense Advanced Research Porjects Agency (DARPA) in 1996, to designate a project for integrating the spin of carriers into semiconductors for device applications (Wolf *et al.*, 2006). Originally, the term was an acronym of "spin transport electronics" but is often used to refer to any combination of transporting spin and electronics. While many authors use the term spintronics for ferromagnetic structures, in this book, we restrict its use to magnetic Si-based materials, in particular by doping TMEs in Si to create half-metals (HMs).

1.1.2 *Spintronic devices*

While conventional electronics use charge carriers to transmit, manipulate and store information, spintronic devices use either spin or both spin and charge for these tasks. The addition of spin in device applications presents many attractive features. Conventional electronic charge carriers have limited speed and suffer from dissipation due to heat loss. A spintronic device may require far less power to operate than its electronic counterpart and do not necessarily require significant sustained power to maintain its magnetic state. Furthermore, an important feature is the non-volatility. Memory modules can maintain their information state while power is turned off (Daughton, 1992).

An ideal class of materials to use for spintronic applications are HMs. HMs have a metallic spin channel and an insulating spin channel, thus their currents are 100% spin polarized. Unfortunately, these types of materials are difficult to grow and remain elusive in technological applications because they can suffer from structural transitions, spin-flip transitions if the Fermi level is near the conduction states of the insulating channel, and the spin–orbit effect, which invalidates spin as an acceptable quantum number.

1.1.2.1 *Spin Polarization*

To understand why half-metals (HMs) are ideal spintronic materials, it is important to understand the origins of the spin polarization and magnetoresistance (MR).

The spin polarization P describes the imbalance of spin states ($\uparrow$ and $\downarrow$) at the Fermi energy (E_F)

$$P = \frac{d_\uparrow - d_\downarrow}{d_\uparrow + d_\downarrow}, \tag{1.1}$$

where $d_\uparrow$ and $d_\downarrow$ are the respective density of states (DOS) of spin up ($\uparrow$) and spin down ($\downarrow$) channels at E_F. The spin polarization can also be expressed in terms of the number of available states $N_\uparrow$ and $N_\downarrow$ in the respective spin channel,

$$P = \frac{N_\uparrow - N_\downarrow}{N_\uparrow + N_\downarrow}. \tag{1.2}$$

The first definition makes it possible to directly determine P from the DOS. Typical ferromagnetic materials have a spin polarization of $P \approx 40$ to 50% at room temperature (RT) (Fe has $P = 45\%$ at RT). A unique feature of HMs is the spin polarization $P = 1$ because E_F falls in the gap of one of the spin channels.

1.1.2.2 *Jullière formula*

MR describes how the resistivity of a material depends on the orientation or strength of the magnetic moment. According to the Jullière formula (Julliere, 1975), the MR is directly related to P at E_F:

$$MR = \frac{2P}{1 - P^2}. \tag{1.3}$$

The MR can be infinite if $P = 1$, a condition that is only satisfied by HMs.

1.1.3 *Applications*

At this point, no device has successfully employed, in their operation, half-metallic materials. Important material issues concerning design, growth, and interfaces to these spintronic materials are crucial to technological progress. If they can be grown and used to fabricate spintronic devices, the applications can parallel the progress made of materials currently found in GMR devices. Other applications of spintronics include spin-based field-effect transistors (FETs), magnetic memory, and quantum computing components based on the spin degree of freedom.

1.1.3.1 *Giant magnetoresistive device*

Typical GMR devices have non-magnetic materials layered between two FM slabs and current flows perpendicular to the plane. This setup is the so-called current perpendicular to plane (CPP) heterostructure and schematically shown in Fig. 1.1. The fundamental operation of the device relies on an imbalance in the DOS near the Fermi surface, or spin polarization P, between the two (antiparallel and parallel) spin orientations. Nonferromagnetic materials, such as pure Si, have spin degeneracy so the electron energies are independent of the electron spin and the occupation of $\uparrow$ and

$\downarrow$ states is equal. Equal occupation of the spin channels results in no spin polarization, $P = 0$. In a FM material, such as iron (Fe), due to the exchange interaction, the spin degeneracy is broken leading to more electrons occupying one spin channel (the spin majority channel) than the other spin channel (the spin minority channel), resulting in a net spin polarizarion, $P \neq 0$. The DOS at E_F determines the number of conducting carriers. When both ferromagnetic layers in a CPP-type device have their magnetic moments oriented parallel, the current of one spin can flow through both layers more easily because these states are available to conduct charge. When the layers have their magnetic moments aligned antiparallel, the current flows easily through one spin channel in one layer, but meets resistance through the other spin channel in the other layer.

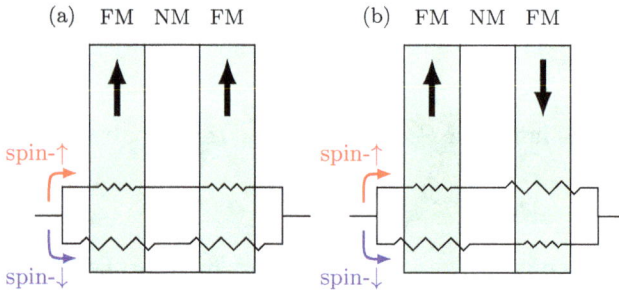

Fig. 1.1 GMR devices in the CPP configuration. Two FM layers sandwich a non-magnetic (NM) layer. The left FM layer has fixed magnetic moment orientation while the right FM layer is allowed to change direction. The size of the resistor symbol denotes the relative resistance of the spin channel in each FM layer. (a) With parallel alignment, spin-up current flows more easily and the overall resistance is low. (b) The two FM layers have antiparallel moments, both spin channels experience high resistance in one of the two layers, thus the overall resistance is high.

1.2 Crystals

Even though the models of spintronic devices are not necessarily perfect crystals, it is possible to describe the doped semiconductors using the same terminology because the doping is typically limited to around 10 %. This section introduces the models and terminology used to discuss crystals, the determination of crystal structures and the electron wave function.

The model of a perfect crystal is a collection of atoms arranged in a definite and repeated fashion, or a lattice. The set of points with identical environment are defined as lattice points of a Bravais lattice. It is therefore possible to (i) define a set of basis vectors (a_1, a_2, and a_3 in 3 dimensions) which translates a point, which may or may not coincide with an atom, to a repeated point, and (ii) define a "unit cell", using the basis vectors, inside which all of the atoms are arranged in a specific pattern.

1.2.1   *Unit cells*

The unit cell is a volume that is repeated throughout the crystal. It is constructed such that the cell can be translated by any number of the basis vectors away and preserve the symmetry of the crystal,

$$\boldsymbol{R} = n_1 \boldsymbol{a}_1 + n_2 \boldsymbol{a}_2 + n_3 \boldsymbol{a}_3 \qquad (1.4)$$

where n_i, $i = 1, 2, 3$ are arbitrary integers. The vector $\boldsymbol{R}$ is called a lattice vector and connects a lattice point to another lattice point. In three dimensions, the three independent basis vectors (a_1, a_2 and a_3), which may or may not lie along the Cartesian coordinate axes, span the unit cell. The unit cell and basis vectors are not unique for any crystals since $a_i \rightarrow n a_i$, where n is an integer, is still considered a basis vector of a larger unit cell (a supercell).

The atoms within a unit cell are called a basis. For example, the NaCl crystal can be defined with 2 atoms, Na and Cl, per unit cell. The positions of the basis are specified by vectors τ_i with respect to the chosen origin. The number of nearest neighbors of an atom is called the coordination number. A two dimensional representation of a section of a crystal is shown in Fig. 1.2.

All of the properties of a solid are determined by the arrangement of the atoms in the unit cell, so it is no longer necessary to deal with 10^{23} atoms. Additionally, the symmetry properties of a crystal are completely specified by the arrangements of the atoms inside the unit cell.

1.2.1.1   *Bravais lattice*

The set of all lattice points defined by Eq. 1.4 is called the Bravais lattice. Each point in the Bravais lattice has an identical environment surrounding it. In three dimensions, there are 14 possible Bravais lattices represented in Fig. 1.3.

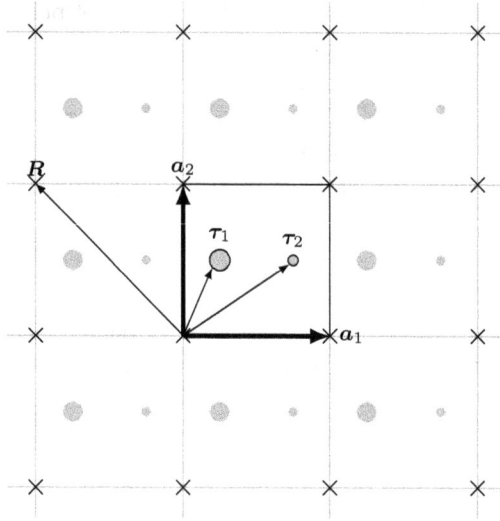

Fig. 1.2   Two dimensional square unit cell with a basis of two atoms. The unit cell is outlined in black, the basis vectors a_1 and a_2 are shown as thick vectors, lattice points are shown as X's, and the basis, described by vectors τ_1 and τ_2. The vector $R = -a_1 + a_2$ denotes a lattice vector and the repeated unit cells and basis atoms are shown in grey. The coordination number is 2 because each atom has a nearest neighbor to the left and the right.

1.2.1.2   *The primitive cell*

When the basis vectors a_1, a_2, a_3 are the shortest vectors that generate the lattice, the three vectors define the primitive cell. The primitive cell is the smallest repeating unit of the crystal, however, there is no unique way to construct the primitive cell.

1.2.1.3   *The Wigner–Seitz cell*

Wigner and Seitz (1933) described a method to generate a primitive cell. The process is

(1) Draw vectors from one lattice point to all its nearest neighbor points.
(2) Bisect the vectors using perpendicular planes.
(3) The volume enclosed by the planes gives the WS primitive cell.

This method is also used when applied to the unit cell in k-space. This is the Brillouin zone, which will be discussed in section 1.2.1.8. The process in 2 dimensions is illustrated in Fig. 1.4.

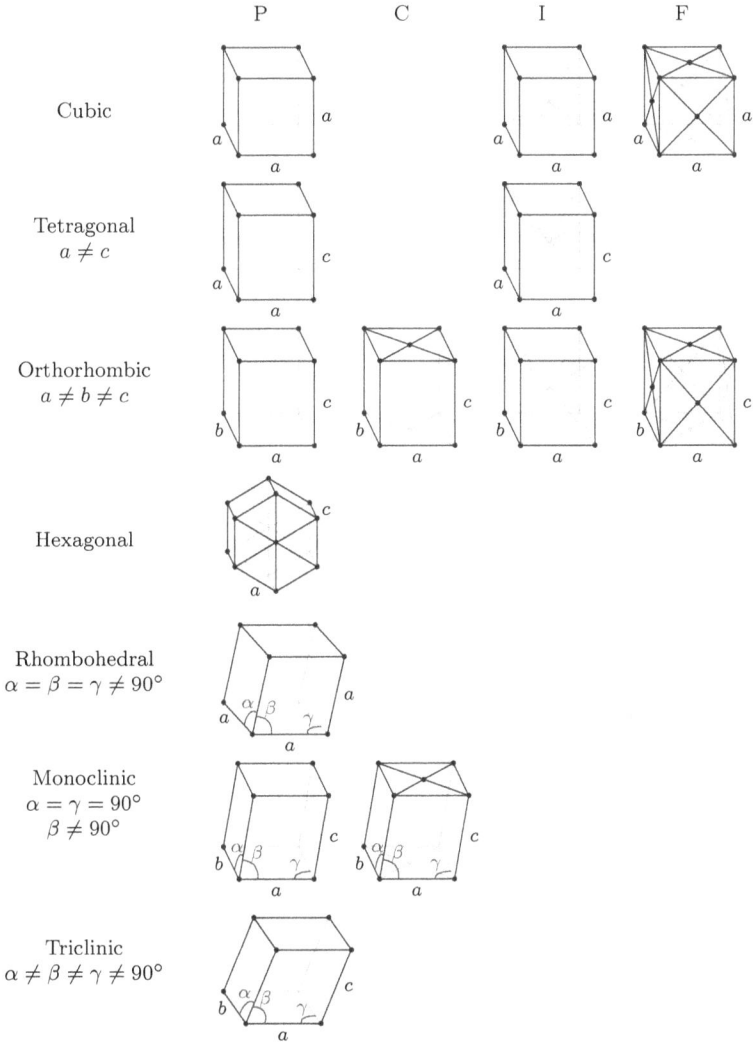

Fig. 1.3   14 Bravais lattices grouped by the 7 lattice systems. The four classes of lattices are primitive (P), base-centered (C), body-centered (I) and face-centered (F).

1.2.1.4   *The conventional cells*

The rectangular cuboid cell is the simplest cell and very convenient to model, so it is useful to define a set of basis vectors where the three vectors are at right angles to each other. This unit cell is called the conventional

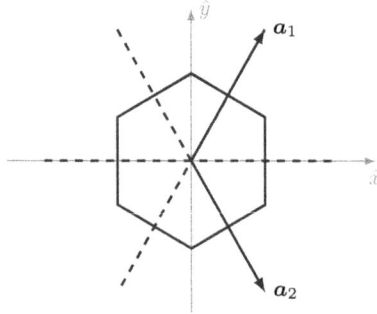

Fig. 1.4   A 2 dimensional WS cell constructed from basis vectors a_1 and a_2. Dashed lines show the possible translations from the origin bisected with connected line segments. The connected line segments (the hexagon) is the perimeter of the WS cell.

cell and often stacked to create "supercells", discussed in the next section. For an example of a conventional cell, consider the body-centered cubic lattice with one atom per cell with the primitive cell defined by

$$a_1 = \frac{a}{2}(-1, 1, 1) \tag{1.5}$$

$$a_2 = \frac{a}{2}(1, -1, 1) \tag{1.6}$$

$$a_3 = \frac{a}{2}(1, 1, -1) \tag{1.7}$$

where the numbers in parenthesis are the components along the $\hat{x}$, $\hat{y}$ and $\hat{z}$ directions, respectively. The conventional cell can be constructed as

$$a_1' = a_2 + a_3 = a(1, 0, 0) \tag{1.8}$$

$$a_2' = a_3 + a_1 = a(0, 1, 0) \tag{1.9}$$

$$a_3' = a_1 + a_2 = a(0, 0, 1) \tag{1.10}$$

and contain two atoms per conventional cell.

1.2.1.5   *Supercells*

For treating Si-based spintronic materials, most of the materials are in alloy forms. To simulate an alloy of the form $M_x Si_{1-x}$, where M is the doped element, a supercell is necessary. A supercell is constructed by stacking unit cells together and then making a change to the basis such that the larger cell becomes the supercell. Conventional cells are often used for this purpose because they are easy to visualize how they stack.

One application of the supercell is for modeling dopant elements into a silicon crystal. For $x \approx 1.6\%$ doping of phosphorus (P) in Si, one way to model this is to stack $2^3 = 8$ conventional cells of Si (each conventional cell of Si has 8 Si atoms) to form a larger cube. Next, one Si atom in the 64 atom cube is replaced by the dopant element.

Antiferromagnetism is also modeled using the supercell technique if there is only one magnetic atom per primitive cell. A primitive cell with one TM atom can only be simulated as a FM material since every magnetic moment vector is repeated on the lattice. By stacking two primitive cells to form a supercell, an antiferromagnetic configuration can be obtained by aligning the two moments of the magnetic atoms antiparallel.

Another useful application of supercells is to create large heterostructures, surfaces and layered structures. In chapter 3, we discuss a few magnetic structures constructed by starting with large supercells of Si.

1.2.1.6 *Periodic functions*

The condition of a periodic lattice causes observables to have the same periodicity as the lattice unless acted upon by an external force. The observables are described by functions that have the same periodicity as the lattice, such as the electronic charge density, must have the same value when the coordinate system is translated by a lattice vector $\boldsymbol{R}$

$$f(\boldsymbol{r} + \boldsymbol{R}) = f(\boldsymbol{r}). \tag{1.11}$$

The Fourier decomposition of the generic function f is

$$\sum_{\boldsymbol{G}} e^{-i\boldsymbol{G}\cdot(\boldsymbol{r}+\boldsymbol{R})} \tilde{f}(\boldsymbol{G}) = \sum_{\boldsymbol{G}} e^{-i\boldsymbol{G}\cdot\boldsymbol{r}} \tilde{f}(\boldsymbol{G}) \tag{1.12}$$

which requires $\exp(i\boldsymbol{G} \cdot \boldsymbol{R}) = 1$, or

$$\boldsymbol{G} \cdot \boldsymbol{R} = 2\pi n \tag{1.13}$$

where n is an integer. The above condition is important because, due to the periodicity, it is often simpler to examine electronic structure properties in reciprocal space, where there is a reciprocal lattice corresponding to the real space lattice. The set of vectors $\boldsymbol{G}$ that satisfy Eq. 1.13 are called reciprocal lattice vectors (RLVs). We suggest imagining any RLV in terms of a wave in real space with wavelength $\lambda = 2\pi/|\boldsymbol{G}|$.

1.2.1.7 *The reciprocal lattice*

The set of RLVs that satisfy Eq. 1.13 defines the reciprocal lattice. In the same way the unit cell in real space forms the crystal lattice, the reciprocal lattice contains repeating unit volumes in reciprocal space, which is often denoted k-space.

Linear integer combinations of the basis vectors a_1, a_2 and a_3 form the lattice vectors R. Similarly, a different set of basis vectors in reciprocal space b_1, b_2, b_3 combine to give the lattice vector G

$$G = m_1 b_1 + m_2 b_2 + m_3 b_3 \qquad (1.14)$$

where m_i are integers. The usual definition for the vectors b_j is

$$a_i \cdot b_j = 2\pi \delta_{ij}, \qquad (1.15)$$

which follows from Eq. 1.13. From the dot product rules, each reciprocal lattice basis vector can be obtained from the basis vectors in real space,

$$b_1 = 2\pi \frac{a_2 \times a_3}{a_1 \cdot a_2 \times a_3} \qquad (1.16)$$

$$b_2 = 2\pi \frac{a_3 \times a_1}{a_2 \cdot a_3 \times a_1} \qquad (1.17)$$

$$b_3 = 2\pi \frac{a_1 \times a_2}{a_3 \cdot a_1 \times a_2}. \qquad (1.18)$$

The basis vectors b_i have unit of 1/length and their linear combinations span the reciprocal space.

1.2.1.8   *The first Brillouin zone*

The vectors b_j form a lattice in reciprocal space. The unit cell in reciprocal space that is constructed using the Wigner–Sietz procedure is called the first Brillouin zone (BZ). As previously suggested, each k-point in the first BZ is associated with a wave in real space. In one dimension we show a wave corresponding to a k-vector within the BZ, $k \leq \frac{\pi}{a}$, in Fig. 1.5 (a). In Fig 1.5 (b), we show a wave corresponding to a k-vector on the zone boundary $k = \frac{\pi}{a}$. In Fig 1.5 (c), a wave with wavelength smaller than the lattice constant is shown in comparison to the wave at the zone boundary.

It is important to have a clear understanding of the first BZ because all other k points in reciprocal space are related to ones in the first BZ. They all have wave lengths larger than a and cannot be used to describe any variations of physical quantities within the unit cell in real space. As a simple example, let us consider the atomic displacement of atoms due to a sound wave within a sample. For the sample, we will use a one-dimensional unit cell with length a with a basis of one atom located at $r = 0$. For the sound wave, we use the wave in Fig. 1.5(b) to describe the y-displacement of atoms at positions $x = \ldots, -a, -a/2, 0, a/2, a, \ldots$. Atoms located at

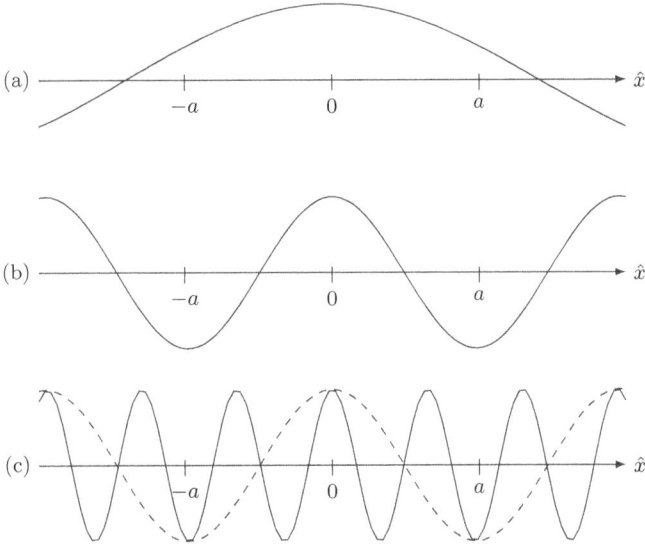

Fig. 1.5   Types of waves in a crystal. (a) wave larger than the unit cell, (b) wave with $\lambda = 2a$, and (c) a wave (solid curve) smaller than the unit cell superimposed with a wave (dashed curve) that describes the same atomic displacements.

integer increments of a displace alternately in the $+y$ or $-y$ directions. The atom displacement is also described by the solid wave in Fig. 1.5(c). The wavelengths of the two waves are $\lambda_{(b)} = 2a$ (dashed curve) and $\lambda_{(c)} = 2a/3$ (solid curve), or $\boldsymbol{k}_{(b)} = \pi/a\hat{\boldsymbol{x}}$ and $\boldsymbol{k}_{(c)} = 3\pi/a\hat{\boldsymbol{x}}$. The vector $\boldsymbol{G}$ can be any integer multiple of $\boldsymbol{b} = 2\pi/a$, so we find that

$$\boldsymbol{k}_{(c)} = \boldsymbol{k}_{(b)} + \boldsymbol{b}. \tag{1.19}$$

The two waves describe the same physical phenomena—the vibration amplitude of the atom at a lattice point—so only the wave within the first BZ is meaningful. The oscillation of the wave between 0 and a does not have any physical meaning since there is no atom between 0 and a. In general, all waves outside the first BZ are related to ones within the first BZ by RLVs.

The similarity of zones in the reciprocal space is also clearly seen in Laue's theory for determining the crystal structure and in Bloch's theorem. In sections 1.2.2 and 1.2.3, we explore these two important aspects of crystals.

1.2.2   *Determination of the crystal structure*

Before discussing Laue's theory, it is important to understand the basic experiments for probing crystal structures. To determine a crystal structure, we are looking for information about the separation between the atoms. The separation of the atoms is of the order of 5 Å. The probe should be able to generate interference on the same length scale. Possible candidates include

- X-rays have energy $E = hc/\lambda$ of about keV so the wave length is ~ 6 Å.
- Thermal neutrons have energy $E \sim 0.1\,\text{eV}$ and mass $m = 10 \times 10^{-24}$ g. The deBroglie wave length is ~ 1 Å.
- Electrons have energy $E \sim 100\,\text{eV}$ and mass $m = 0.9 \times 10^{-27}$ g, so the de Broglie wave length is also ~ 1 Å.

Although neutrons and electrons are within the correct wave length scale, they can also suffer from other scattering, which can obscure the measurements especially for magnetic materials as the neutron has a magnetic moment. Electrons, in particular, are charged so they can scatter if there is a charged dopant. An x-ray is the best candidate for the probe. An x-ray incident on a crystal will scatter against the electron cloud surrounding atoms in a plane as depicted in Fig. 1.6 and the interference pattern can be measured.

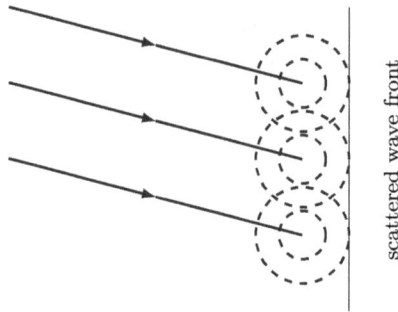

Fig. 1.6   Waves scattering from atoms in a plane.

1.2.2.1   *Miller indices*

The orientation of planes of atoms that x-rays scatter from are called Miller indices. They are used to specify vectors and planes in the crystal. Vectors are denoted as $[hjk]$, where h, j, and k are integer multiples of the basis vectors $\boldsymbol{a}_1$, $\boldsymbol{a}_2$ and $\boldsymbol{a}_3$:

$$[hjk] \to h\boldsymbol{a}_1 + j\boldsymbol{a}_2 + k\boldsymbol{a}_3. \tag{1.20}$$

When a component of the vector is negative, a "bar" is placed over the index. For example, the vector $\boldsymbol{r} = -5\boldsymbol{a}_1 + 2\boldsymbol{a}_3$ is denoted as $[\bar{5}02]$

The description for planes is a bit more complicated. The notation (hjk) is used to describe a plane that intersects the axis at distances that are proportional to the inverse of the indices h, j and k. In other words, the vector $[hjk]$ is normal to the plane. The plane $(2\bar{1}1)$, for example, is a plane that contains the three points $\boldsymbol{a}_1/2$, $-\boldsymbol{a}_2$ and $\boldsymbol{a}_3$. This plane is shown in Fig. 1.7(a). If one or two indices are zero, such as (120), then the plane is parallel to those basis directions. In Fig. 1.7(b), we show the plane (120) where the plane does not intercepts the $\boldsymbol{a}_3$ axis.

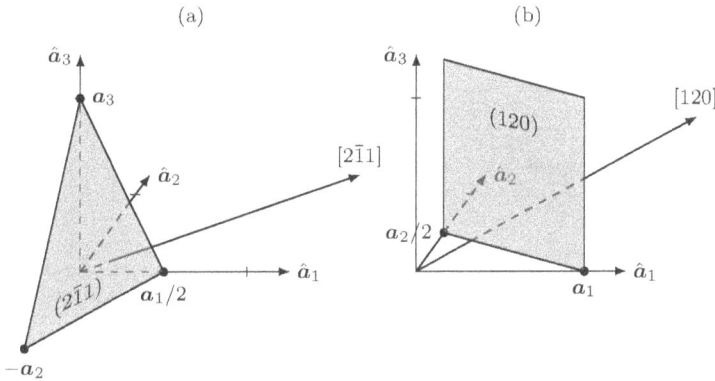

Fig. 1.7 Examples of planes described by Miller indices. (a) $(2\bar{1}1)$. (b) (120)

1.2.2.2 *X-ray diffraction*

The detailed process of x-ray diffraction is that an x-ray in incident on the electrons around the ions, the electrons polarize which, in turn, produce dipoles and re-radiate the x-ray. This process is very much like Huygen's principle in optics illustrated in Fig. 1.6. The interference of the scattered waves gives the diffraction pattern with constructive interference when the path length difference between rays scattered from adjacent atoms is an integral multiple of the x-ray wavelength λ.

The simplest x-ray method is the Bragg's law reflection over parallel planes, described by Miller indices, inside the crystal. The path difference between waves reflected off of different planes result in the diffraction pattern. Bragg's law, as described in most elementary physics books is illustrated in Fig. 1.8. There is constructive interference when the path difference $2d \sin \theta$ is an integer multiple of the x-ray wavelength

$$n\lambda = 2d \sin \theta \qquad (1.21)$$

where d is the distance between parallel planes described by Miller indices.

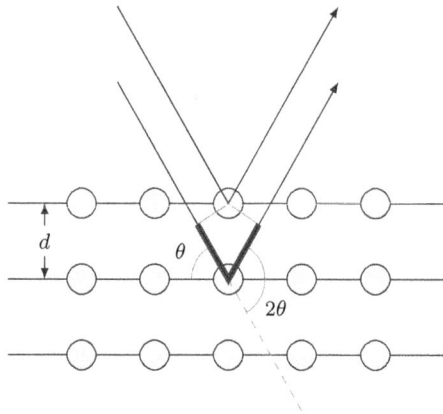

Fig. 1.8   Schematic of Bragg's law. X-rays incident on the planes of atoms scatter, travel different path lengths and interfere, forming an interference pattern. The path difference between the two x-rays is highlighted by the thick black line.

1.2.2.3   *Laue's Theory*

Bragg's law suffers from two major limitations: (1) the atoms are treated as points and (2) the scattered wave is treated as a plane wave. In reality, the electrons are located some distance away from the ion centers and the scattered wave propagates radially away from the scattering center. Laue proposed the following changes to the model:

- the crystal is treated as an array of points in a macroscopically small, but microscopically large region,
- the incident wave is treated as a uniform plane wave, and
- the scattered wave is treated as a spherical wave radiated from a certain point of the array.

The incident and scattered waves are

$$F(\boldsymbol{r}) = F_0 e^{i\boldsymbol{k}\cdot\boldsymbol{r}} \tag{1.22}$$

$$F_s(r) = F_0 e^{i\boldsymbol{k}\cdot\boldsymbol{\rho}}\frac{1}{r}e^{ikr}, \tag{1.23}$$

respectively, where we assume the plane wave hits the electron cloud surrounding the atom at $\boldsymbol{\rho}$ with respect to an arbitrarily chosen origin and the magnitude of the incoming $\boldsymbol{k}$ and outgoing $\boldsymbol{k}'$ waves are identical. The spherical wave emanates from position $\boldsymbol{\rho}$, so $\boldsymbol{r}$ is measured from $\boldsymbol{\rho}$ to the detector. This geometry is shown in Fig. 1.9.

The atom at $\boldsymbol{\rho}$ is a microscopic distance from the origin while the detector is a macroscopic distance from the origin. Using $R \gg \rho$,

$$r = R\left(1 + \frac{\rho^2}{R^2} - 2\hat{\boldsymbol{R}}\cdot\boldsymbol{\rho}\right)^{1/2} \tag{1.24}$$

$$\approx R\left(1 - \frac{\hat{\boldsymbol{R}}}{R}\cdot\boldsymbol{\rho}\right) \tag{1.25}$$

to leading order in ρ, where $\hat{\boldsymbol{R}}$ is the unit vector in the direction of the detector. The denominator in $F_s(r)$ can be simply replaced by $r = R$ since the exponent is much more sensitive to small changes in r

$$F_s = \frac{F_0}{R}e^{ikR}e^{i(\boldsymbol{k}\cdot\boldsymbol{\rho}-k\hat{\boldsymbol{R}}\cdot\boldsymbol{\rho})} \tag{1.26}$$

$$= \frac{F_0}{R}e^{ikR}e^{i\Delta\boldsymbol{k}\cdot\boldsymbol{\rho}} \tag{1.27}$$

with $\Delta\boldsymbol{k} = \boldsymbol{k} - \boldsymbol{k}'$.

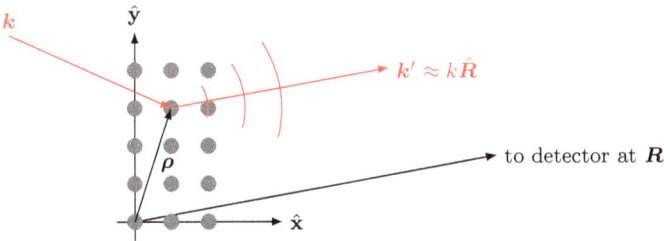

Fig. 1.9 Schematic of the geometry in Laue's theory. Incoming x-ray $\boldsymbol{k}$ scatters from the electron cloud around the atom at $\boldsymbol{\rho}$ towards the detector at $\boldsymbol{R}$. Since the $\boldsymbol{R}$ is much larger than $\boldsymbol{\rho}$, the wave scattered to the detector is in the same approximate direction as $\boldsymbol{R}$, or $\boldsymbol{k}' \approx k\hat{\boldsymbol{R}}$.

In the simplest scenario, the spherical wave scattered from each lattice point at

$$\boldsymbol{\rho} = n_1\boldsymbol{a}_1 + n_2\boldsymbol{a}_2 + n_3\boldsymbol{a}_3 \tag{1.28}$$

so the interference pattern is a sum of terms from each lattice point

$$I = I_0 \sum_{n_1,n_2,n_3} e^{i\boldsymbol{\rho}\cdot(\Delta\boldsymbol{k})} \qquad (1.29)$$

where $\Delta\boldsymbol{k} = \boldsymbol{k} - k\hat{\boldsymbol{R}}$. The contribution from each lattice vector is separable so the intensity at the detector is

$$I^2 = I_0^2 \left|\sum_{n_1} e^{in_1 \boldsymbol{a}_1 \cdot \Delta\boldsymbol{k}}\right|^2 \left|\sum_{n_2} e^{in_2 \boldsymbol{a}_2 \cdot \Delta\boldsymbol{k}}\right|^2 \left|\sum_{n_3} e^{in_3 \boldsymbol{a}_3 \cdot \Delta\boldsymbol{k}}\right|^2 \qquad (1.30)$$

The summations are

$$\left|\sum_{n_1} e^{in_1 \boldsymbol{a}_1 \cdot \Delta\boldsymbol{k}}\right|^2 = \frac{\sin^2 N_1 \boldsymbol{a}_1 \cdot \Delta\boldsymbol{k}}{\sin^2 \boldsymbol{a}_1 \cdot \Delta\boldsymbol{k}} \qquad (1.31)$$

where N_1 is the number of scattering centers in the $\boldsymbol{a}_1$ direction. The summation is a maximum when $\boldsymbol{a}_1 \cdot \Delta\boldsymbol{k} = 2\pi p_1$, where p_1 is an integer. Similarly, $\boldsymbol{a}_2 \cdot \Delta\boldsymbol{k} = 2\pi p_2$ and $\boldsymbol{a}_3 \cdot \Delta\boldsymbol{k} = 2\pi p_3$. These three conditions are called the Laue conditions and are satisfied when

$$\Delta\boldsymbol{k} = \boldsymbol{G}, \qquad (1.32)$$

a RLV from Eqs. 1.16–1.18. This expression relates the incident plane wave $\boldsymbol{k}$ to a scattered plane wave $\boldsymbol{k}'$ in every other BZ including the first BZ by a RLV. Laue's conditions are a generalization of Bragg's Law in reciprocal space.

1.2.3   Bloch's theorem

The periodicity of the crystal also influences the symmetry of the electron states through Schrödinger's equation, so we are interested in the forms of the wave functions that the electrons can have in a crystal. Bloch's theorem (Bloch, 1929) provides an important solution to electron wave function. In the single-particle model, an electron moves in a crystal experiencing a periodic potential from the ions

$$U(\boldsymbol{r} + \boldsymbol{R}) = U(\boldsymbol{r}) \qquad (1.33)$$

so the Hamiltonian is

$$\hat{H} = \frac{1}{2m}\hat{p}^2 + U(\boldsymbol{r}) \qquad (1.34)$$

where $\hat{p}$ is the momentum operator. Due to the periodicity of the potential, Bloch (1929) argued that the active lattice translation transformation

$$T_{\boldsymbol{R}} = \mathrm{e}^{-i\hat{p}\cdot\boldsymbol{R}} \qquad (1.35)$$

commutes with the Hamiltonian. Therefore the solutions to the Hamiltonian Ψ_n must also be eigenvectors of the translation operator

$$\hat{H}\Psi_n(r) = E_n\Psi_n(r) \tag{1.36}$$

$$\hat{T}_R\Psi_n(r) = \Psi_n(r - R) \tag{1.37}$$

$$= t_R\Psi_n(r) \tag{1.38}$$

where n is the band index (a complete set of single particle states in the unit cell),

$$t_R = e^{-i\boldsymbol{k}\cdot\boldsymbol{R}} \tag{1.39}$$

are eigenvalues of the translation operator, and $\boldsymbol{k}$ is the wave vector associated with the momentum $\boldsymbol{p} = \hbar\boldsymbol{k}$. The symmetry of the potential leads to the symmetry of the wave function

$$\Psi_{n\boldsymbol{k}}(\boldsymbol{r} + \boldsymbol{R}) = e^{i\boldsymbol{k}\cdot\boldsymbol{R}}\Psi_{n\boldsymbol{k}}(\boldsymbol{r}). \tag{1.40}$$

Equation 1.40 is Bloch's theorem. It is important to note that the wave function depends on $\boldsymbol{k}$.

1.2.3.1 *Bloch waves*

An alternate form of Bloch's theorem separates the wave function into two parts, a plane wave part and a periodic part

$$u_{n\boldsymbol{k}}(\boldsymbol{r} + \boldsymbol{R}) = u_{n\boldsymbol{k}}(\boldsymbol{r}). \tag{1.41}$$

A Bloch wave is defined as

$$\Psi_{n\boldsymbol{k}}(\boldsymbol{r}) = e^{i\boldsymbol{k}\cdot\boldsymbol{r}}u_{n\boldsymbol{k}}(\boldsymbol{r}). \tag{1.42}$$

A comparison between the Bloch wave and the periodic part of the Bloch wave is depicted in Fig. 1.10. In (a), the $\boldsymbol{k}$ wave is within the first BZ. In (b), the periodic part of the wave function has the same periodicity as the unit cell. When combined in (c), the full Bloch wave (the solid line) is shown as the combination of the two parts in (a) and (b).

1.2.3.2 *Zone folding*

For an arbitrary wave function solution for the point $\boldsymbol{k}'$ in reciprocal space, the function can be "folded" back to the first BZ as discussed in section 1.2.1.8. A RLV can be chosen such that $\boldsymbol{k} = \boldsymbol{k}' - \boldsymbol{G}$ lies within the first BZ. This property of Bloch waves is called "zone-folding" since the

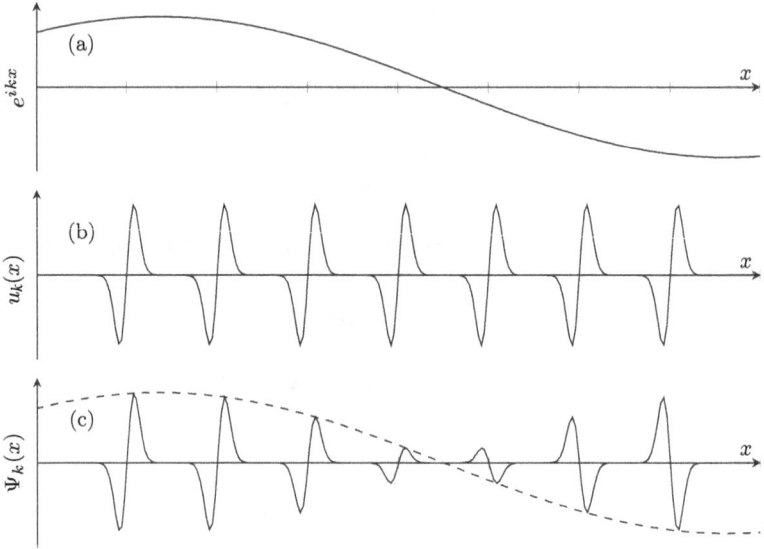

Fig. 1.10   A Bloch wave in one dimension. (a) The wave within the first BZ showing the envelope function e^{ikx}. (b) The periodic function $u_{nk}(x)$ with the periodicity of the unit cell. (c) The complete Bloch wave (solid line) and its envelope function (dashed curve) with wave number k.

zones outside the first BZ are folded back into the first BZ. The relationship between wave functions at the two point is

$$\Psi_{nk'}(\boldsymbol{r}) = \Psi_{nk+G}(\boldsymbol{r}) \tag{1.43}$$

$$= u_{nk+G}(\boldsymbol{r})e^{i(k+G)\cdot r} \tag{1.44}$$

$$= \left[u_{nk+G}(\boldsymbol{r})e^{iG\cdot r}\right]e^{ik\cdot r}. \tag{1.45}$$

The term in the bracket is also a periodic part of a Bloch wave so

$$\left[u_{nk+G}(\boldsymbol{r})e^{iG\cdot r}\right]e^{ik\cdot r} = u'_{nk}e^{ik\cdot r} \tag{1.46}$$

is also a solution to the Schrödinger equation.

1.2.4   *Modulated crystal structures*

In section 3.5, we will discuss a type of magnetic material with a modulated crystal structure. For some of these structures, such as non-magnetic Na_2CO_3 (De Wolff, 1974) and some magnetic Mn doped Si (Miyazaki *et al.*,

2008), the measured x-ray diffraction (XRD) cannot be satisfied by three RLVs in the Laue condition. Instead, another vector b_4 has to be included

$$G = m_1 b_1 + m_2 b_2 + m_3 b_3 + m_4 b_4 \qquad (1.47)$$

where b_4 is in the same 3-dimensional space, so

$$b_4 = \alpha_1 b_1 + \alpha_2 b_2 + \alpha_3 b_3 \qquad (1.48)$$

where at least one of the α_i are irrational. These crystals can be represented as having some periodic property, or modulation, that is not commensurate with the unit cell that defines the RLVs b_1, b_2 and b_3. Examples of such properties that can be incommensurate with the unit cell exist in magnetic spiral structures, atom displacements, atomic substitutions, or any combination of the above.

One method to model these modulated systems, described by De Wolff (1974), involves constructing a supercell that approximately makes the periodicity of the modulation commensurate with the unit cell. For example, a helical modulation of the atom position in the $\hat{x}$-direction with wavelength of $\lambda = \sqrt{2}a_1$, where a_1 is the unit cell lattice constant in the $\hat{x}$-direction, could be approximately represented by stacking 5 unit cells and fitting 7 wavelengths inside, as shown in Fig. 1.11.

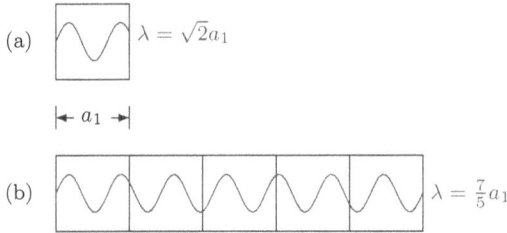

Fig. 1.11   Modulated waves within unit cells. The incommensurate wave (a) is approximated by the supercell (b) with rational wavelength compared to the unit cell length.

1.3   Spin dependent interactions

Magnetism in condensed matter is due to the alignment of orbital and spin moments of the electron. There are several interactions that cause the spin moments of electrons to either align or anti-align. The direct exchange is the most basic interaction because it involves the correlation of spins directly through the exchange interaction: a force resulting from the

Coulomb interaction and the quantum mechanical properties of fermions, such as the Pauli principle. Spins may also interact indirectly in several ways. The double-exchange interaction involves itinerant spins with varying valencies occupying an intermediary vacancy in a non-magnetic element valence state and favors ferromagnetism. Superexchange similarly involves an intermediate element, but the spins interact through direct-exchange with the valence states of the intermediate element. Finally, in the Ruderman–Kittel–Kasuya–Yosida (RKKY) interaction (Ruderman and Kittel, 1954; Kasuya, 1956; Yosida, 1957), two local spins interact mediated by conduction electrons. Since HMs contain localized moments and a metallic channel, the only indirect interaction we will discuss is the RKKY interaction.

For TMEs, which are used to dope semiconductors, the magnetism is primarily caused by the spins of the d-electrons. In FM materials, the exchange interaction is responsible for the imbalance of spin occupation. The exchange interaction is strong enough, such as in the case of Fe, to cause many of the electrons spontaneously align along the same direction, thus occupying the same kind of spin state. Above a threshold temperature, called the Curie temperature (T_C), the thermal energy overcomes the exchange interaction and the magnetic system become paramagnetic.

In this section, we focus on the basic mechanism of direct-exchange. Next, we discuss the RKKY interaction as an example of an indirect interaction. Finally, we discuss the spin–orbit (S–O) interaction which couples the spin with the orbital motion of the electrons.

1.3.1   *Direct fermion exchange*

The imbalance of spin-state occupation in FM materials is due to the Pauli exclusion principle and the Coulomb interaction. Without the Pauli exclusion effect, the magnetic dipole interaction between the moments of electrons could cause the spins on neighboring atoms to align antiparallel, however its $1/r_{ij}^3$ dependence, where r_{ij} is the separation between magnetic dipoles at i and j, makes it extremely weak compared to the measured exchange interaction. The exchange interaction arises as two electrons approach each other and their orbital wave functions overlap differently depending on their spin orientations. When the spins are aligned antiparallel, shown in Fig. 1.12(a), the Pauli exclusion principle demands the orbital part of the wave function be symmetric. This allows the electron densities to get close giving a large Coulomb energy. However, the coulomb energy can be reduced greatly if one spin flips so the spins are parallel and the orbital wave function is antisymmetric, shown in Fig. 1.12(b). The electrons are not allowed to get near each other and the Coulomb energy is lowered. This causes the electrons to have parallel spins states and gives ferromagnetism.

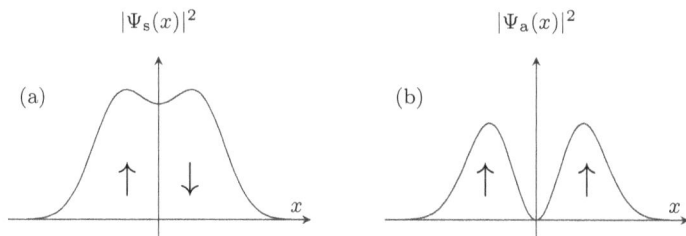

Fig. 1.12 A simple schematic diagram of the exchange interaction. (a) Two-particle density with antisymmetric spin and symmetric orbital wave function. The orbitals can get near each other in the center region and the Coulomb energy is higher. (b) Two-particle density with symmetric spin and antisymmetric orbital wave function. The orbitals separate and the Coulomb energy is lower.

1.3.1.1 *Simple two-electron system*

In order to model the fermionic exchange interaction, it is instructive to examine the situation of two electrons in a subshell of an atom, such as $3d^2$, in more detail. We will start with the simple case that each electron can occupy two distinct orbitals ϕ_a and ϕ_b described by the independent one-particle Hamiltonians $\hat{H}_1$ and $\hat{H}_2$

$$\hat{H}_1\phi_a(\boldsymbol{r}_1) = \epsilon_a\phi_a(\boldsymbol{r}_1) \tag{1.49}$$

$$\hat{H}_2\phi_b(\boldsymbol{r}_2) = \epsilon_b\phi_b(\boldsymbol{r}_2) \tag{1.50}$$

with spin states α and β. Most commonly, α and β are described as spin-up ($\uparrow$) and spin-down ($\downarrow$) states. We account for the Pauli exclusion principle by combining these states using the Slater determinant. Four distinct two-particle trial wave functions arise that are antisymmetric under particle exchange. They are

$$\psi_{\alpha\alpha}(\boldsymbol{r}_1, \boldsymbol{r}_2) = \frac{1}{\sqrt{2}}\alpha(1)\alpha(2)\left[\phi_a(\boldsymbol{r}_1)\phi_b(\boldsymbol{r}_2) - \phi_a(\boldsymbol{r}_2)\phi_b(\boldsymbol{r}_1)\right] \tag{1.51}$$

$$\psi_{\beta\beta}(\boldsymbol{r}_1, \boldsymbol{r}_2) = \frac{1}{\sqrt{2}}\beta(1)\beta(2)\left[\phi_a(\boldsymbol{r}_1)\phi_b(\boldsymbol{r}_2) - \phi_a(\boldsymbol{r}_2)\phi_b(\boldsymbol{r}_1)\right] \tag{1.52}$$

$$\psi_{\alpha\beta}(\boldsymbol{r}_1, \boldsymbol{r}_2) = \frac{1}{\sqrt{2}}\left[\phi_a(\boldsymbol{r}_1)\phi_b(\boldsymbol{r}_2)\beta(1)\alpha(2) - \phi_a(\boldsymbol{r}_2)\phi_b(\boldsymbol{r}_1)\beta(2)\alpha(1)\right] \tag{1.53}$$

$$\psi_{\beta\alpha}(\boldsymbol{r}_1, \boldsymbol{r}_2) = \frac{1}{\sqrt{2}}\left[\phi_a(\boldsymbol{r}_1)\phi_b(\boldsymbol{r}_2)\alpha(1)\beta(2) - \phi_a(\boldsymbol{r}_2)\phi_b(\boldsymbol{r}_1)\alpha(2)\beta(1)\right] \tag{1.54}$$

where the first two describe states that are parallel spin states and the last two are antiparallel spin configurations.

So far, the two particles are not interacting. The interaction between the two electrons is the Coulomb potential,

$$\hat{H} = \hat{H}_1 + \hat{H}_2 + \hat{V} \tag{1.55}$$

$$\hat{V} = \frac{e^2}{|\boldsymbol{r}_1 - \boldsymbol{r}_2|} \tag{1.56}$$

and the ground state can be determined from the variational principle. Using this method, the two-particle states can be described by a linear combination of two-particle basis states. Using the four noninteracting particle states as a basis, the ground state can be written

$$\Psi(\boldsymbol{r}_1, \boldsymbol{r}_2) = \sum_i c_i \psi_i \tag{1.57}$$

where the i is the set of $\alpha\alpha$, $\beta\beta$, $\alpha\beta$ and $\beta\alpha$, and c_i are values that will be varied. The expectation value of the Hamiltonian is

$$\left\langle \Psi \left| \hat{H} \right| \Psi \right\rangle = \sum_{ij} c_i^* c_j \left\langle \psi_i | \hat{H} | \psi_j \right\rangle. \tag{1.58}$$

The two-particle wave functions are orthonormal, so the values of c_i that give the minimum energy is equivalent to solving the matrix eigenvalue problem

$$\sum_j \left\langle \psi_i \left| \hat{H} \right| \psi_j \right\rangle c_j = \epsilon c_i \tag{1.59}$$

$$Hc = \epsilon c, \tag{1.60}$$

where H is the 4×4 matrix, c is a 4×1 eigenvector and ϵ is the eigenvalue. The matrix H is

$$H = E_0 + \begin{pmatrix} K - J & 0 & 0 & 0 \\ 0 & K - J & 0 & 0 \\ 0 & 0 & K & -J \\ 0 & 0 & -J & K \end{pmatrix} \tag{1.61}$$

where $E_0 = (\epsilon_a + \epsilon_b)I$ is the non interacting energies from Eqs. 1.49 and 1.50, I is the identity matrix, K is the Coulomb integral and J is the positive definite exchange integral

$$K = \frac{1}{2} \iint dr_1 dr_2 \frac{|\phi_a(\boldsymbol{r}_1)|^2 |\phi_b(\boldsymbol{r}_2)|^2}{|\boldsymbol{r}_1 - \boldsymbol{r}_2|} \tag{1.62}$$

$$J = \frac{1}{2} \iint dr_1 dr_2 \frac{\phi_a^*(\boldsymbol{r}_1)\phi_b(\boldsymbol{r}_1)\phi_b^*(\boldsymbol{r}_2)\phi_a(\boldsymbol{r}_2)}{|\boldsymbol{r}_1 - \boldsymbol{r}_2|}. \tag{1.63}$$

It is important to note that in J, ϕ_a depends on $\boldsymbol{r}_2$ and ϕ_b depends on $\boldsymbol{r}_1$ to account for the particle exchange. The parallel spin terms in the Hamiltonian are diagonal with energy $E = E_0 + K - J$ and total spin $S = 1$, where we have set $\hbar = 1$, so all that is left is the 2×2 matrix for the oppositely oriented spin configurations. The energies for these states are $E = E_0 + K \pm J$. The lower energy $E = E_0 + K - J$ is due to the state comprised of the sum of spin states

$$|10\rangle = \frac{1}{\sqrt{2}} (\psi_{\alpha\beta}(\boldsymbol{r}_1, \boldsymbol{r}_2) + \psi_{\beta\alpha}(\boldsymbol{r}_1, \boldsymbol{r}_2)) \tag{1.64}$$

$$= \frac{1}{2} (\phi_a(\boldsymbol{r}_1)\phi_b(\boldsymbol{r}_2) - \phi_a(\boldsymbol{r}_2)\phi_b(\boldsymbol{r}_1)) (\alpha(1)\beta(2) + \alpha(2)\beta(1)) \tag{1.65}$$

with symmetric combination of spin states, thus the total spin is $S = 1$ with $S_z = 0$. The last state is the difference of the oppositely oriented spin states with energy $E = E_0 + K + J$

$$|00\rangle = \frac{1}{\sqrt{2}} (\psi_{\alpha\beta}(\boldsymbol{r}_1, \boldsymbol{r}_2) - \psi_{\beta\alpha}(\boldsymbol{r}_1, \boldsymbol{r}_2)) \tag{1.66}$$

$$= \frac{1}{2} (\phi_a(\boldsymbol{r}_1)\phi_b(\boldsymbol{r}_2) + \phi_a(\boldsymbol{r}_2)\phi_b(\boldsymbol{r}_1)) (\alpha(1)\beta(2) - \alpha(2)\beta(1)) \tag{1.67}$$

and has an antisymmetrtic spin state, thus $S = 0$ and $S_z = 0$. The three degenerate states with energy $E = E_0 + K - J$ are called the spin triplet states and demonstrate that the parallel alignment of spins has lower energy than the antiparallel alignment due to direct exchange. The $E = E_0 + K + J$ state has higher energy and is called the spin singlet state.

The spin operators $S^2 = (\boldsymbol{S}_1 + \boldsymbol{S}_2)^2$, S_1^2 and S_2^2 all commute with the Hamiltonian, so they share the eigenvectors associated with the singlet and triplet states. These operators are related by the dot product

$$\boldsymbol{S}_1 \cdot \boldsymbol{S}_2 = \frac{1}{2} \left(S^2 - S_1^2 - S_2^2 \right) \tag{1.68}$$

$$= \frac{1}{2} [s(s+1) - s_1(s_1+1) - s_2(s_2+1)] \tag{1.69}$$

$$= \begin{cases} 1/4 & \text{for spin triplet states} \\ -3/4 & \text{for spin singlet state.} \end{cases} \tag{1.70}$$

Using the two values provided by the dot product, the energy of the two particle system can be written

$$E = E_0 + K - \frac{1}{2}J - 2J\boldsymbol{S}_1 \cdot \boldsymbol{S}_2 \tag{1.71}$$

$$= \begin{cases} E_0 + K - J & \text{for spin triplet states} \\ E_0 + K + J & \text{for spin singlet state.} \end{cases} \tag{1.72}$$

1.3.1.2 *Hartree–Fock approximation*

When the number of electrons N is much greater than 2, we turn to the Hartree–Fock approximation of the exchange energy. In the independent particle model, N fermions are described by N single-particle wave functions $\psi_{\sigma i}(x)$ where σ is the spin index ($\uparrow$ or $\downarrow$) and $i = 1 \ldots N$. The simplest approximation to the many-body wave function is to expand the wave function into a product of single-particle wave functions $\phi_i(\boldsymbol{r}_i)$. The direct product of wave functions is

$$\Psi_{\mathrm{DP}}(x_1, \ldots, x_N) = \phi_1(x_1)\phi_2(x_2)\cdots\phi_N(x_N). \tag{1.73}$$

The expectation value of the Coulomb energy using this approximation is

$$\left\langle \Psi_{\mathrm{DP}} \left| \frac{1}{|x_i - x_j|} \right| \Psi_{\mathrm{DP}} \right\rangle = \frac{1}{2}\sum_{ij} \iint dxdx' \frac{|\psi_i(x)|^2|\psi_j(x')|^2}{|x - x'|} \tag{1.74}$$

$$= U_{\mathrm{H}} \tag{1.75}$$

which overestimates the Coulomb energy because the direct product does not correctly capture the exchange symmetry of the wave function due to Pauli's exclusion principle. The correct symmetry should have the property

$$\Psi(\ldots, x_i, \ldots, x_j, \ldots) = -\Psi(\ldots, x_j, \ldots, x_i, \ldots) \tag{1.76}$$

for any i, j pair that is exchanged. One choice for the wave function that satisfies the property in Eq. 1.76 is the Slater determinant

$$\Psi_{\mathrm{SD}} = \begin{vmatrix} \psi_1(x_1) & \psi_1(x_2) & \cdots & \psi_1(x_N) \\ \psi_2(x_1) & \psi_2(x_2) & \cdots & \psi_2(x_N) \\ \vdots & \vdots & \ddots & \vdots \\ \psi_N(x_1) & \psi_N(x_2) & \cdots & \psi_N(x_N) \end{vmatrix}. \tag{1.77}$$

Carrying out the Coulomb term using the Slater determinant results in a second term that is subtracted from the Hartree term

$$\left\langle \Psi_{\text{SD}} \left| \frac{1}{|x_i - x_j|} \right| \Psi_{\text{SD}} \right\rangle = \frac{1}{2} \sum_{ij} \iint dx dx' \frac{|\psi_i(x)|^2 |\psi_j(x')|^2}{|x - x'|} \tag{1.78}$$

$$- \frac{1}{2} \sum_{ij} \delta_{\sigma_i \sigma_j} \iint dx dx' \frac{\psi_j^*(x') \psi_i(x') \psi_i^*(x) \psi_j(x)}{|x - x'|}. \tag{1.79}$$

The Kronecker delta function for the spins assures that only parallel spin contributions are removed from the Hartree term. The subtracted term is referred to as the Fock, or exchange, integral. The Coulomb energy in this form is called the Hartree–Fock energy.

1.3.1.3 *The Heisenberg model*

The Heisenberg model of electron spins is widely used to treat interacting, localized electron spins due to the exchange interaction. Based on the dot product of spin operators in the two-electron case, the Hamiltonian for the Heisenberg model is

$$\hat{H} = -\frac{1}{2} \sum_{\langle ij \rangle} J_{ij} \boldsymbol{S}_i \cdot \boldsymbol{S}_j \tag{1.80}$$

where the notation $\sum_{\langle ij \rangle}$ means the summation is done over nearest neighboring localized spins $\boldsymbol{S}_i$ and $\boldsymbol{S}_j$ only, J_{ij} is the strength of the interaction between spins i and j, and the factor of $1/2$ accounts for double-counting the pairs. For positive values of J_{ij}, the Heisenberg Hamiltonian prefers parallel, or ferromagnetic, configuration of neighboring spins due to the dot product. If J_{ij} is negative, on the other hand, antiparallel spin alignment between neighbors is preferred, giving rise to antiferromagnetism.

1.3.2 *The RKKY interaction*

The RKKY interaction indirectly couples the localized magnetic moment of TMEs through the polarization of conduction electrons over a long range. This interaction is important in spintronic materials, especially HMs, because they are metallic and their large magnetic moments are primarily localized.

Ruderman and Kittel (1954) first investigated the indirect exchange between nuclear spins $\boldsymbol{I}_i$ and $\boldsymbol{I}_j$ by the hyperfine interaction with the spin $\boldsymbol{S}$ of conduction electrons in a periodic lattice. Similarly, Kasuya (1956) considered the exchange interaction between conduction electrons, approximated by Bloch wave functions, and 1/2-filled localized atomic states centered around R_i that contribute spin S_i to the local magnetic moment. In a simple picture (Yosida, 1957), the spin of the localized atomic state S_i polarizes the electrons near the Fermi surface and induces a magnetization density in the conduction band

$$M_c(r - R_i) \sim S_i. \tag{1.81}$$

The magnetization of the conduction band then interacts with the localized spin at another site S_j through the Zeeman-type interaction

$$U(R_{ij}) \sim M_c(R_{ij}) \cdot S_j \tag{1.82}$$

where $R_{ij} = R_j - R_i$ is the distance between localized moments S_j and S_i. The net effect is that the energy is dependent on the spin orientations.

$$U = J_{\text{RKKY}}(R_{ij}) S_i \cdot S_j \tag{1.83}$$

where J_{RKKY} is the indirect coupling strength of the RKKY interaction. Unlike the Heisenberg exchange, where the interaction is usually limited to nearest neighbor interactions, the RKKY interaction is long ranged. The strength of the interaction J_{RKKY} oscillates as a function of R_{ij}, as shown in Fig. 1.13. The oscillation of J_{RKKY} gives rise to either ferromagnetic or antiferromagnetic alignment of the spins.

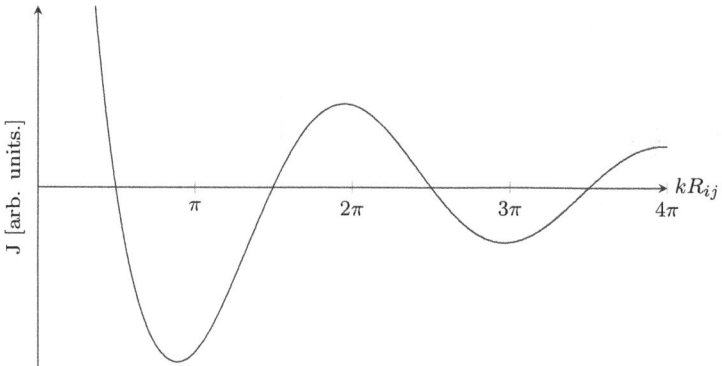

Fig. 1.13   The strength of the RKKY interaction as a function of distance between localized spins.

1.3.3   The spin–orbit interaction

Coupling of the spin and orbital angular momenta arises from the relativistic effect of the orbital parts of the motion on the spin part. The interaction is

$$H_{\text{so}} = \xi \boldsymbol{L} \cdot \boldsymbol{S} \tag{1.84}$$

where ξ is the strength of the S–O interaction, $\boldsymbol{L}$ is the orbital angular momentum operator and $\boldsymbol{S}$ is the spin moment operator. The simplest way to see how an electron's spin interacts with its own angular motion is to consider an electron, represented in the Bohr model, moving with velocity $\boldsymbol{v} = \boldsymbol{p}/m_e$ around the nucleus providing the central potential $U(\boldsymbol{r})$. In the electron frame of reference, the nucleus is moving with velocity $-\boldsymbol{v}$, effectively behaving as a current loop, and generates a magnetic field

$$\boldsymbol{B} = \frac{1}{m_e e c^2} \frac{1}{r} \frac{\partial U(r)}{\partial r} \boldsymbol{L} \tag{1.85}$$

where $\boldsymbol{L} = \boldsymbol{r} \times \boldsymbol{p}$ is the angular momentum of the electron in the lab frame. The spin of the electron contributes the magnetic moment

$$\boldsymbol{\mu}_e = -\frac{g \mu_B}{\hbar} \boldsymbol{S}, \tag{1.86}$$

where $g \approx 2$ is the g-factor, so the energy splitting associated with the orbital motion of the electron with its spin moment should be

$$E_{\text{so}} = -\boldsymbol{\mu}_e \cdot \boldsymbol{B} \tag{1.87}$$

$$= \frac{2\mu_B}{\hbar m_e e c^2} \frac{1}{r} \frac{\partial U(r)}{\partial r} \boldsymbol{L} \cdot \boldsymbol{S} \tag{1.88}$$

$$\xi = \frac{2\mu_B}{\hbar m_e e c^2} \frac{1}{r} \frac{\partial U(r)}{\partial r} \tag{1.89}$$

The value in Eq. 1.89 is, however, exactly twice as large as experimentally determined values of the splitting. Thomas (1926) attributed this additional contribution to the acceleration of the electron reference frame. The additional relativistic contribution causes the electron spin to precess around the magnetic field, as shown in Fig. 1.14, with the Thomas precession rate (Jackson, 1998)

$$\boldsymbol{\Omega}_T = -\frac{\gamma^2}{\gamma + 1} \frac{\boldsymbol{a} \times \boldsymbol{v}}{c^2}, \tag{1.90}$$

where γ is the usual relativistic boost, c is the speed of light, $\boldsymbol{v}$ is the velocity of the electron in the lab frame and $\boldsymbol{a}$ is the acceleration of the electron in the lab frame. The real value of the S–O interaction strength is

$$\xi = \frac{\mu_B}{\hbar m_e e c^2} \frac{1}{r} \frac{\partial U(r)}{\partial r} \tag{1.91}$$

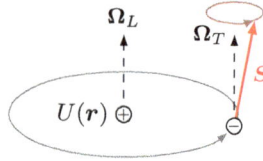

Fig. 1.14   The relativistic motion of the electron spin $\boldsymbol{S}$ in the lab frame.

The S–O interaction is important when a material involves a heavy atom with a large principal quantum number. Elements with 4d electrons and some with 3d electrons should have significant S–O coupling. When S–O coupling is introduced, neither the spin S nor the orbital L angular momenta are good quantum numbers since the wave function can no longer be decomposed into spin and orbital subspaces. Instead, the total angular momentum quantum number J is used as a good quantum number and involves the sum of both spin types.

1.4   Half-metals

Now that we understand how the electron spin behaves, we can discuss the concept of HMs. Earlier, we mentioned that HMs are ideal materials for spintronic applications. In the following, we discuss the history and the development of HMs. de Groot *et al.* (1983) first predicted that the half-Heusler alloy NiMnSb would exhibit 100 % spin polarization because the the ↓ states act metallic while the ↑ states act insulating. The E_F is located within the gap of the ↑ states and intersects the ↓ states. The DOS at E_F of the ↓ states is finite while it is zero for the ↑ states. According to Eq. 1.3, $P = 1$. A necessary condition that arises from 100 % spin polarization, and when the g-factor is 2, is that the magnetic moment of the sample is an integer. This new material property is coined "half-metallicity."

1.4.1   *Conditions for half-metals*

From a theoretical approach, it is possible to formulate two conditions on HMs. The first condition is that one of its spin channels exhibits metallic behavior while the oppositely oriented spin channel shows semiconducting or insulating properties. A schematic energy diagram of a HM is shown in Fig. 1.15(a). E_F intersects the finite DOS of the majority-spin channel and is located in the gap of the minority-spin states. For comparison, the energy diagram for the states of a FM metal is shown in Fig. 1.15(b) where E_F intersects both spin channels. For a HM $P = 1$, while P of iron, for example, is 0.45. In the perfect situation of $P = 1$, MR $\to \infty$. Consequently, HMs are ideal candidates for spintronic devices that are used in for MR.

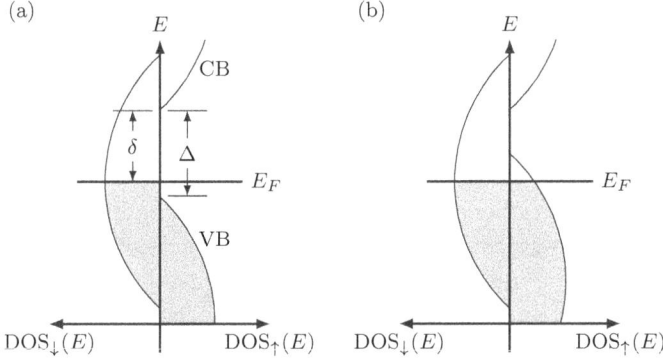

Fig. 1.15 (a) Schematic of a half-metallic spin-polarized energy diagram showing the conduction band and the valence band. E_F cuts though a non-zero DOS in the down-spin channel and the gap in the up-spin channel. Δ indicates the fundamental gap in the up-spin channel and δ denotes the spin-flip gap. Another gap, not shown in the figure, exists $\xi = \Delta - \delta$ from the top of the $\uparrow$ VB to the $\downarrow$ E_F. (b) Schematic of a FM energy diagram. Spin channels have unequal, but non-zero DOS at E_F in the two spin channels.

The second condition is that the magnetic moment per unit cell should be an integer. This second condition should be strictly satisfied by any theoretical calculations. The reasons are:

(1) The total number of electrons per unit cell N_t is an integer.
(2) The number of electrons per unit cell filling up to the top of the VB in the insulating channel, N_v is an integer. The moments of these electrons are canceled by an equal number of electrons in the metallic spin channel. There are $2N_v$ electrons not contributing to the magnetic moment in the sample.
(3) The number of electrons contributing to the magnetic properties is $N_t - 2N_v$, an integer.
(4) For electrons in 3d metals, such as Mn, the g-factor is 2 due to negligible spin-orbit (SO) interaction, so the magnetic moment is one Bohr magneton μ_B per contributing electron per unit cell.
(5) The magnetic moment per unit cell is $N_t - 2N_v$, an integer, in units of Bohr magneton μ_B.

1.4.2   *Interactions resulting in half-metallicity*

In Heusler alloys and several zincblende (ZB) compounds, such as MnC and MnAs, consist of at least one TME with partially filled 3d-shells and a group IV or V element with a full s-shell and partially filled p-shell in valence. The TME also contains a filled s-shell which simply transfer to the group IV or V element. Both classes of materials have similar environments around the TME so we may use ZB as a prototype.

1.4.2.1   *Crystal field*

Under the tetrahedral crystal field of the ZB structure, the 5 d-states split into two sets, one doubly-degenerate set with lower energy called e_g states (d_{z^2} and $d_{x^2-y^2}$) and one triply-degenerate set having higher energy called t_{2g} (d_{xy}, d_{yz}, and d_{zx}) states (Tinkham, 2003). The states are depicted in Fig. 1.16. The e_g states have lobes pointing towards the second neighbors and do not form bonds with the neighboring non-metal element (and are hence called non-bonding states). The t_{2g} orbitals have higher energy because their orbital lobes are closer to the electrons surrounding neighboring elements in the tetrahedral environment.

higher energy   lower energy

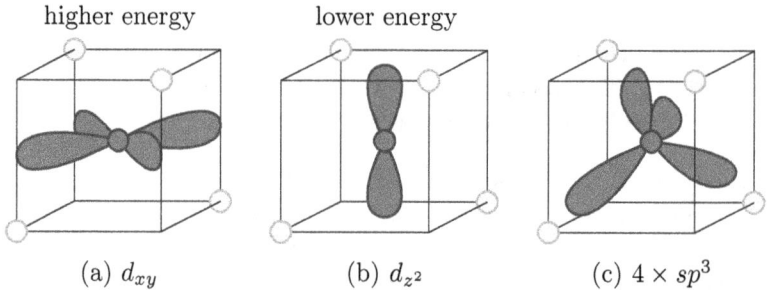

(a) d_{xy} (b) d_{z^2} (c) $4 \times sp^3$

Fig. 1.16   Electron orbitals (circles) in the tetrahedral environment of 4 elements (circles). (a) The d_{xy}-type orbital has four lobes that are near the neighbors. Linear combinations of the d_{xy}-type orbitals point towards the neighbors. (b) The d_{z^2} orbital has two lobes which are between two sets of neighbors. (c) The four sp^3 orbitals point directly towards the neighboring elements.

Similarly, the s and p electrons, of group IV or V elements, hybridize to form four degenerate sp^3 orbitals in the tetrahedral environment that also point to neighboring atoms.

1.4.2.2 *d-p hybridization*

The t_{2g} and sp^3 orbitals overlap along the tetrahedral directions. The hybridization of the two types of states create bonding states with lower energy and anti-bonding states with higher energy. Their energy difference forms a bonding-antibonding gap $\Delta_{bond-antibond}$. Fig. 1.17 shows the entire interaction schematically. At this point, the ↑ and ↓ states are degenerate.

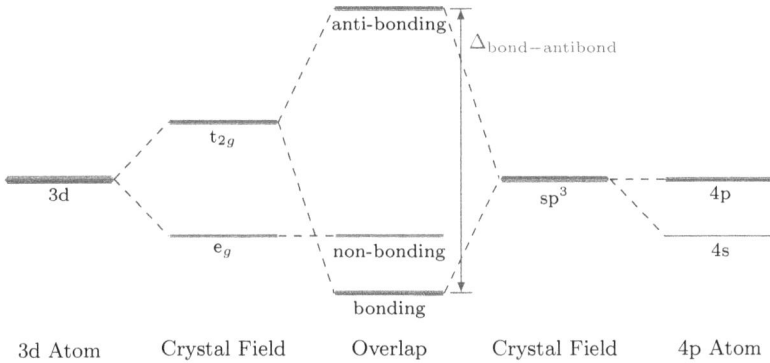

Fig. 1.17 Schematic drawing showing the splitting of atomic states and formation of bonding-antibonding states in a ZB crystal. The vertical direction denotes an arbitrary energy scale. The 3d states split from the crystal field into e_g and t_{2g} states. The 3s and 3p states hybridize to form sp^3 orbitals. Finally, the sp^3 and t_{2g} form bonding and antibonding states with a gap of $\Delta_{bond-antibond}$. This is called d-p hybridization.

1.4.2.3 *Exchange interactions*

The exchange interaction between the electrons shifts the electrons from anti-parallel spin states to parallel spin states and lowers the total energy of the system. When the three interactions properly compete, E_F intersects states of one spin channel and falls in the gap of the oppositely oriented spin states.

1.4.3 *Estimating the magnetic moment*

1.4.3.1 *Slater–Pauling rules*

Slater (1936) and Pauling (1938) independently tried to explain the decrease in saturation magnetic moments from Fe ($2.2\,\mu_B$) to Ni ($0.6\,\mu_B$). Slater also considered alloys of Fe–Ni and proposed the first plot of M versus atomic numbers (Slater, 1936). Later, contributors developed the plot into the empirical Slater-Pauling curve Chikazumi (1964) and Bozorth (1968).

Using Fe as a reference, it turns out the alloys of magnetic TME, forming local moments, can be located to the left of Fe while Ni and Co, with moments contributed by both local moments (3d-orbitals) and itinerant electrons (hybridized $3d^2 4s4p^2$-orbitals), are at the right of Fe, shown in Fig. 1.18. Pauling argued that the decrease in the magnetic moment on the right side of Fe is due to the increased number of localized d-electrons anti-aligning their spins.

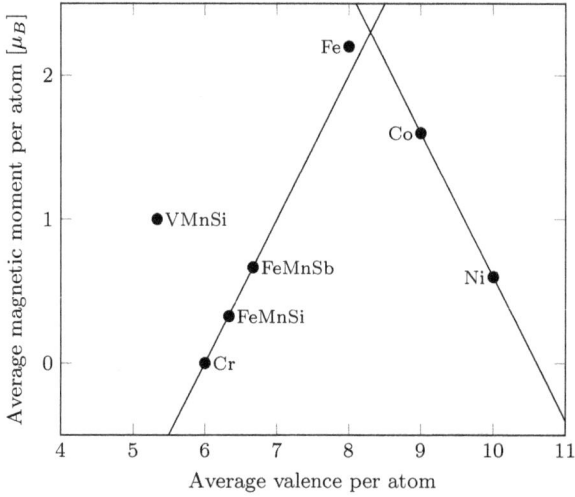

Fig. 1.18   The Slater–Pauling curve. The number of valance electrons and the magnetic moment are divided by the number of atoms per unit cell so that alloys and Cr are on the same curve.

Kübler (1984) focused only on the local moment part of the Slater–Pauling curve. He suggested to use the calculated number of bands in the insulating channel N_{insul} as a parameter to determine the moment of HMs. The moment is

$$M = N - 2N_{\text{insul}} \tag{1.92}$$

in units of μ_B, where N is the total number of occupied bands. The g-factor of 2 for the 3d electrons is assumed. The rule predicts the moment well—agreeing with the results of calculations if M is a positive integer. However, this rule does not account for any interaction causing half-metallicity so it's predictions do not agree if M is negative (Shaughnessy *et al.*, 2013). We suggest that if the rule is used judiciously; it is useful to select elements for growing new compounds that could exhibit half-metallic properties but it does not provide any predictive insight for new classes of compounds.

1.4.3.2 *The ionic model*

There is another way to predict the moment of a HM. Schwarz (1986) was the first to predict CrO_2 should have an integer magnetic moment, however let us use CrAs as an example. The As element will accept three electrons from Cr. This leaves three electrons remaining at the Cr atom to form a local moment by aligning their spin due to Hund's first rule. With the g-factor of 2, the moment is $3\,\mu_B$ and agrees with the calculation (Pask *et al.*, 2003).

1.4.4 *Mechanisms that hinder half-metallicity*

1.4.4.1 *Spin–orbit (S–O) interactions*

The states of a system involving SO interactions cannot be thought of as purely up- or down-spin states, so the notion of a HM involving SO coupling becomes unclear. The definition of a HM relies on spin being a good quantum number. At E_F, the DOS must have one spin channel occupied and no available states in the other spin channel. If the spin states of a system with SO coupling are mixed, how can there be HMs?

One prominent feature of HMs is the integer magnetic moment resulting from E_F in the gap of the insulating channel. When the SO interaction is introduced, it is possible that the magnetic moment remains an integer. While finding an integer magnetic moment does not guarantee that the material is a HM, it can be a strong indication that the SO interaction does not have a large effect.

An alternate view of SO interactions in potential HMs is to apply perturbation theory. Using the spin-polarized solution without SO interactions as the unperturbed solutions, the $\uparrow$ and $\downarrow$ spin states are still good quantum numbers. Pickett and Eschrig (2007) investigated the issue of SO coupling in HMs using perturbation theory on a model system. They based the unperturbed system on a model HM involving d-electrons: three degenerate t_{2g} states, each with spin $\uparrow$ or $\downarrow$ giving 6 states total. They ignored the e_g states because, in many ZB HMs, they are too far in energy from E_F. They assumed the $\uparrow$ and $\downarrow$ states were split by the exchange parameter Δ. The perturbing spin–orbit Hamiltonian is

$$H_{\text{so}} = \frac{\xi}{2}\left(2L_z S_z + L_+ S_- + L_- S_+\right) \tag{1.93}$$

where $L_\pm = L_x \pm iL_y$ and $S_\pm = S_x \pm iS_y$ are the transverse angular momentum and spin-moment operators, respectively. The term ξ is defined for the electron as

$$\xi(r) = \frac{\mu_B}{\hbar m_e c^2} \left(\frac{1}{r} \frac{dV}{dr} \right) \tag{1.94}$$

and is related to the average electric field (dV/dr) experienced by the electron. The perturbing Hamiltonian in Eq. 1.93 contains spin raising and lowering operators so some spin mixing $(DOS_\uparrow / DOS_\downarrow)$ near E_F is anticipated.

Using the model system with ξ and Δ as parameters, Pickett and Eschrig found that the degree of mixing, when $\xi \ll \Delta$, is on the order of $(\xi/\Delta)^2$. Here, we will demonstrate that the same factor of mixing arises for more general systems. For simplicity, we will also assume that ξ is a small parameter. In numerical calculations, perturbation theory can be used to determine the amount of spin mixing that contributes to the magnetic moment. Consider a crystal with one atom per unit cell (at the origin): the Bloch wave functions can be expanded onto the spherical harmonics around the origin

$$\phi_{n\boldsymbol{k}\sigma}(\boldsymbol{r}) = \sum_{\boldsymbol{G}} a_{n\boldsymbol{k}\sigma}(\boldsymbol{G}) \sum_{lm} \frac{i^l}{2l+1} Y_{lm}^*(\boldsymbol{k}+\boldsymbol{G}) Y_{lm}(\boldsymbol{r}) j_l(|\boldsymbol{k}+\boldsymbol{G}|r), \tag{1.95}$$

or in shorter notation,

$$|\phi_{n\boldsymbol{k}\sigma}\rangle = \sum_{lm} |lm\rangle \langle lm | \phi_{n\boldsymbol{k}\sigma}\rangle \tag{1.96}$$

where Y_{lm} are spherical harmonics, j_l are the spherical Bessel functions and σ is either $\uparrow$ or $\downarrow$. Using this form, we can see which terms will appear in the perturbation. The SO Hamiltonian consists of two parts. The first part $(L_z S_z)$ is diagonal

$$\left\langle \phi_{n\boldsymbol{k}\sigma}^{(0)} \left| L_z S_z \right| \phi_{n\boldsymbol{k}\sigma}^{(0)} \right\rangle = \xi \sum_{lm} m\sigma \left| \left\langle \phi_{n\boldsymbol{k}\sigma}^{(0)} | lm \right\rangle \right|^2 \tag{1.97}$$

but does not contribute any spin mixing. The second term involves spin and orbital moment raising and lowering operators

$$\xi \left\langle \phi_{n\boldsymbol{k}\sigma}^{(0)} \left| L_+ S_- + L_- S_+ \right| \phi_{n\boldsymbol{k}\sigma}^{(0)} \right\rangle \tag{1.98}$$

however, this quantity is zero because the spins in the states are the same, thus higher orders of perturbation are required. To first order, the wave function near E_F in the $\downarrow$ spin channel are

$$|\phi_{n\boldsymbol{k}\downarrow}\rangle = |\phi_{n\boldsymbol{k}\downarrow}^{(0)}\rangle + \sum_{n \neq n'} |\phi_{n'\boldsymbol{k}\uparrow}^{(0)}\rangle \frac{\left\langle \phi_{n'\boldsymbol{k}\uparrow}^{(0)} \left| \xi \boldsymbol{L} \cdot \boldsymbol{S} \right| \phi_{n\boldsymbol{k}\downarrow}^{(0)} \right\rangle}{E_{n'\boldsymbol{k}\uparrow} - E_{n\boldsymbol{k}\downarrow}} + \cdots \tag{1.99}$$

The terms involving L_zS_z are zero in the matrix element because of the spin difference. The other term only has L_-S_+ contributing:

$$\left\langle \phi^{(0)}_{n'\boldsymbol{k}\uparrow} \,\middle|\, L_+S_- + L_-S_+ \,\middle|\, \phi^{(0)}_{n\boldsymbol{k}\downarrow} \right\rangle. \tag{1.100}$$

The second order correction to the wave function is

$$\Delta\left(|\phi_{n\boldsymbol{k}\downarrow}\rangle\right) = \xi \sum_{n\neq n'} \frac{|\phi^{(0)}_{n'\boldsymbol{k}\uparrow}\rangle}{E_{n'\boldsymbol{k}\uparrow} - E_{n\boldsymbol{k}\downarrow}}$$
$$\times \sum_{lm} \sqrt{l(l+1)-m(m-1)} \,\langle \phi^{(0)}_{n'\boldsymbol{k}\uparrow}|lm-1\rangle \langle lm|\phi^{(0)}_{n\boldsymbol{k}\downarrow}\rangle \tag{1.101}$$

$$= \xi \sum_{n\neq n'} \frac{|\phi^{(0)}_{n'\boldsymbol{k}\uparrow}\rangle}{E_{n'\boldsymbol{k}\uparrow} - E_{n\boldsymbol{k}\downarrow}} F_{nn'} \tag{1.102}$$

where $F_{nn'}$ is a complex matrix element that simply replaces the sum over l and m. These matrix elements are non-zero and, due to the gap in the $\uparrow$ spin channel, the correction is largest for components where the denominator is on the order of Δ, so the correction is on the order of ξ/Δ.

In SO systems, the magnetic moment of the perturbed system in the z-direction is

$$\sum_{nk\sigma} \langle \phi_{n\boldsymbol{k}\sigma} \,|\, L_z - S_z \,|\, \phi_{n\boldsymbol{k}\sigma} \rangle = M_0 + \Delta M_z \tag{1.103}$$

where M_0 is the magnetic moment of the unperturbed system and ΔM_z is the perturbation correction to the magnetic moment in the z-direction. The lowest non-zero correction is

$$\Delta M_z = \xi^2 \sum_{n'\neq n} \sum_{n''\neq n} \frac{\sum_{lm}(m+1/2)\langle \phi^{(0)}_{n''\boldsymbol{k}\uparrow}|lm\rangle \langle lm|\phi^{(0)}_{n'\boldsymbol{k}\uparrow}\rangle}{(E_{n''\boldsymbol{k}\uparrow} - E_{n\boldsymbol{k}\downarrow})(E_{n'\boldsymbol{k}\uparrow} - E_{n\boldsymbol{k}\downarrow})} F^*_{nn''} F_{nn'}. \tag{1.104}$$

The largest contributing term is when $n' = n''$ and the denominator is on the order of Δ^2, so the correction is proportional to $(\xi/\Delta)^2$.

There are three terms in Eq. 1.104 that contribute to the deviation of the spin magnetic moment from an integer:

(1) The strength of the SO parameter ξ.
(2) The energy denominator describing the difference in energy between spin-$\uparrow$ and spin-$\downarrow$ wave functions.
(3) The mixing of states represented by the term $F_{nn'}$: the matrix element effect.

If all three terms contribute to a negligible contribution to the magnetic moment, then it is reasonable to accept the spin as a good quantum number and describe the half-metallic state.

1.4.4.2 *Thermal spin-flip transitions*

In Fig. 1.15(a), the DOS of a typical HM is shown. Spin-flip transitions may occur from E_F to the bottom of conduction band in the insulating channel if the thermal energy exceeds the spin-flip gap $\Delta - \delta$. These transitions destroy the half-metallicity since both spin channels can conduct current.

The temperature where the onset of thermal spin-flip transitions occurs is denoted T-star (T^*) and is much smaller than T_C. For the HM NiMnSb, $T^* \approx 0.14T_C \approx 80$ to $100\,\mathrm{K}$ (Borca *et al.*, 2001; Hordequin *et al.*, 1997, 1996). Recently, Aldous *et al.* (2012c) grew thin-film MnSb showing $T^* \approx 0.48T_C \approx 350$ to $620\,\mathrm{K}$ (Aldous *et al.*, 2012c). The ranges of values depend on if the theoretical or experimental value of T_C is used.

Chapter 2

Methods of Studying Spintronics

2.1 Theory

Theoretical models and calculations are important for understanding spintronic physics because they allow us to gain insight into the microscopic properties of materials. The modeling is also a powerful tool to design new spintronic materials and to decide whether it is worthwhile to grow and how hard it would be to grow. It is important to use theoretical models that accurately and reliably determine the properties of the electronic system because many physical properties are ultimately tied to the electron behavior. Empirical methods to determine properties are useful, but they rely on parameters that must be chosen post-hoc. Using a variety of models based on physical principles, intuition, and numerical techniques, many material properties tied to an electronic system can now be accurately and reliably calculated particularly using first-principles, or *ab initio*, methods based on density functional theory (DFT) by Hohenberg and Kohn (1964).

The non-relativistic Hamiltonian for a system of positively charged nuclei and negatively charged electrons consists of five terms, two kinetic energies and three Coulomb interaction potentials:

$$H = T_e + T_n + V_{ee} + V_{nn} + V_{ne}, \tag{2.1}$$

where T_e and T_n are the electron kinetic energy and nuclei kinetic energy, respectively, V_{ee} is the electron-electron interaction, V_{nn} is the nuclei-nuclei interaction , and V_{ne} is the electron-nuclei interaction (Marder, 2000). The wave function

$$\Psi(\boldsymbol{r}_1, s_1, \ldots, \boldsymbol{r}_N, s_N, \boldsymbol{R}_1, S_1, \ldots, \boldsymbol{R}_M, S_M, t) \tag{2.2}$$

depends on $4N + 4M + 1$ coordinates where N is the number of electrons and M is the number of nuclei in the system. The number 4 is for each spatial coordinate plus the spin degree of freedom for each particle and the remaining dimension is for the time coordinate.

Multiple approximation techniques have been developed to reduce the many-body problem specified in Eq. 2.1 into manageable forms. The Born–Oppenheimer (BO) approximation (Born and Oppenheimer, 1927), for example, has been employed to separate the motion of ions and electrons. Since the electron mass is so small compared to the nuclear mass the nuclei move slowly compared to the electron motion. The BO approximation is a type of adiabatic approximation because the electrons are instantaneously adjusted according to the positions of the ions. The kinetic energy of the nucleus, T_n can be ignored and V_{nn}, which only depends on the separation distance between nuclei, can be treated as a constant

$$T_e + V_{ee} + V_{ne} = E_e \tag{2.3}$$

where E_e is the total electronic energy of the system. The electronic energy is related to the total energy by

$$E = T_n + V_{nn} + E_e. \tag{2.4}$$

Eq. 2.3 is still a many-body Hamiltonian, but the number of variables is greatly reduced.

Typically, Eq. 2.3 is solved using methods based on DFT, Hartree, or Hartree–Fock approximations which reduce the electronic problem into an effective electron problem. The majority of this section is devoted to solving this equation using DFT. If the motion of the nuclei is important, such as in the case of phonons and other structural changes, the electronic energy E_e may be reintroduced back into Eq. 2.4 to solve for the instantaneous velocity of the nuclei. When the positions of the nuclei change, the electronic energy is no longer valid because V_{ne} depends on the position of the nuclei. Equation 2.3 must be evaluated for each new arrangement of nuclei. In these lecture notes, we are mainly interested in solutions to Eq. 2.3 so we will not discuss phonons, molecular dynamics, or any aspect of physics related to the motion of nuclei.

2.1.1   *Density functional theory*

An important, and popular, foundation for determining material properties is DFT, laid out by Hohenberg and Kohn in the mid-1960s (Hohenberg and Kohn, 1964). This theory uses the desnity ρ, instead of the wave function, as the primary quantity of interest. DFT reduces the complicated many-electron ground state into a problem of determining the optimized electron density.

2.1.1.1 Hoenberg–Kohn Theorems

Fundamental to DFT are the famous Hohenberg–Kohn (HK) theorems (Hohenberg and Kohn, 1964):

Theorem 2.1 (Hohenberg–Kohn 1). *There is a one-to-one correspondence between the ground state density ρ of an N-electron system and the external potential v_{ext} acting on it.*

The *reducio ad absurdum* proof demonstrates that two potentials leading to the same density, must be the same within some additive constant.

Proof. The Hamiltonian of the system is

$$H = T_e + V_{ee} + V_{\text{ext}}, \text{ with} \tag{2.5}$$

$$\langle \Psi \,|\, V_{\text{ext}} \,|\, \Psi \rangle = \int dr \rho(\boldsymbol{r}) v_{\text{ext}}(\boldsymbol{r}) \text{ and} \tag{2.6}$$

$$\rho(\boldsymbol{r}) = \left\langle \Psi(\boldsymbol{r}_1, \ldots, \boldsymbol{r}_N) \left| \sum_{i=1}^{N} \delta(\boldsymbol{r} - \boldsymbol{r}_i) \right| \Psi(\boldsymbol{r}_1, \ldots, \boldsymbol{r}_N) \right\rangle. \tag{2.7}$$

Let Ψ_G be the ground state for the Hamiltonian with v_{ext} that gives density ρ_G. Assume there is another external potential v'_{ext} that gives the same density ρ_G. The two systems are

$$H\Psi_G = (H_0 + V_{\text{ext}})\,\Psi_G = E_G \Psi_G, \text{ and} \tag{2.8}$$

$$H'\Psi'_G = (H_0 + V'_{\text{ext}})\,\Psi'_G = E'_G \Psi'_G. \tag{2.9}$$

The energy of Ψ'_G in the system with V_{ext} is

$$\langle \Psi'_G \,|\, H \,|\, \Psi'_G \rangle = \langle \Psi'_G \,|\, H_0 + V_{\text{ext}} \,|\, \Psi'_G \rangle \tag{2.10}$$

$$= \langle \Psi'_G \,|\, H_0 + V'_{\text{ext}} \,|\, \Psi'_G \rangle + \langle \Psi'_G \,|\, V_{\text{ext}} - V'_{\text{ext}} \,|\, \Psi'_G \rangle \tag{2.11}$$

$$= E'_G + \langle \Psi'_G \,|\, V_{\text{ext}} - V'_{\text{ext}} \,|\, \Psi'_G \rangle \tag{2.12}$$

$$= E'_G + \int dr \rho(\boldsymbol{r})(v_{\text{ext}}(\boldsymbol{r}) - v'_{\text{ext}}(\boldsymbol{r})) \tag{2.13}$$

$$> E_G \tag{2.14}$$

since Ψ'_G is not the ground state of the H system. Similarly,

$$\langle \Psi_G \,|\, H' \,|\, \Psi_G \rangle = E_G + \langle \Psi_G \,|\, V'_{\text{ext}} - V_{\text{ext}} \,|\, \Psi_G \rangle \tag{2.15}$$

$$= E_G + \int dr \rho(\boldsymbol{r})(v'_{\text{ext}}(\boldsymbol{r}) - v_{\text{ext}}(\boldsymbol{r})) \tag{2.16}$$

$$> E'_G \tag{2.17}$$

since Ψ_G is not the ground state of the H' system. By adding the energies of the two systems, we find a contradiction, namely

$$E_G + E'_G < E'_G + \int d\boldsymbol{r}\rho(\boldsymbol{r})(v_{\text{ext}}(\boldsymbol{r}) - v'_{\text{ext}}(\boldsymbol{r})) \tag{2.18}$$

$$+ E_G + \int d\boldsymbol{r}\rho(\boldsymbol{r})(v'_{\text{ext}}(\boldsymbol{r}) - v_{\text{ext}}(\boldsymbol{r})) \tag{2.19}$$

$$E_G + E'_G < E'_G + E_G. \tag{2.20}$$

The assumption that two distinct potentials give the same density is false, so any two potentials that gives the same density must be the same within an additive constant. $\qquad\square$

Theorem 2.2 (Hohenberg–Kohn 2). *The ground state ρ is one that minimizes the total energy functional.*

The total energy is

$$E[\rho] = \langle \Psi \,|\, T_e + V_{ee} \,|\, \Psi \rangle + \int d\boldsymbol{r}\rho(\boldsymbol{r})v_{\text{ext}}(\boldsymbol{r}) \tag{2.21}$$

$$= F[\rho] + V_{\text{ext}}[\rho], \tag{2.22}$$

where F is a functional comprised of the kinetic energy functional T and the interaction energy of electrons V_{ee}. The external potential energy functional V_{ext} is the external potential and often takes the form of the ionic potential. Eq. 2.22 does not account for the first HK theorem. Only the external potential determines the density ρ, but different wave functions that give ρ may have differing expectation values of $\langle \Psi \,|\, T_e + V_{ee} \,|\, \Psi \rangle$. To account for the first HK theorem, F must be redefined so that any Ψ that results in ρ gives the same functional for $F[\rho]$. Levy and Perdew (1985) defined the universal functional using the constrained search

$$F[\rho] = \min_{\Psi \to \rho} \langle \Psi \,|\, T_e + V_{ee} \,|\, \Psi \rangle \tag{2.23}$$

where $\min_{\Psi \to \rho}$ gives the minimum value of the functional on the right using all wave functions Ψ that have density ρ. F is now a universal functional of ρ, that is, it is only dependent on the density and not the form of the wave function.

Proof. This second proof relies on the Euler–Lagrange variational principle using the constraint that the number of particles remain constant

$$N = \int d\boldsymbol{r}\rho(\boldsymbol{r}). \tag{2.24}$$

From the previous theorem, a particular choice of $v_{ext}(r)$ determines a unique ρ, so using the ground state density $\rho_0(r)$ the variation of the total energy with respect to the density is

$$\frac{\delta}{\delta\rho(r)}\left\{F[\rho] + V_{ext}[\rho] - \mu\left(\int dr\rho(r) - N\right)\right\}\bigg|_{\rho=\rho_0} = 0 \qquad (2.25)$$

$$\frac{\delta F[\rho_0]}{\delta\rho_0(r)} + v_{ext}(r) = \mu \qquad (2.26)$$

where μ gives a stationary point. From the definition of in Eq. 2.23, we know that $F[\rho]$ gives the minimum value for the choice of ρ, so we now find that μ must be a minima of the ground state density. Multiplying Eq. 2.26 by ρ and integrating over all r, we find that the total energy is

$$E[\rho_0] = N\mu = E_0 \qquad (2.27)$$

is a minimum of the total energy functional. This proves the second HK theorem. $\qquad\qquad\qquad\qquad\qquad\qquad\qquad\qquad\qquad\qquad\qquad\qquad\square$

Hohenberg and Kohn also demonstrated how the universal DFT functional could be transformed into an expression that separates the classical potential from the exchange-correlation (XC) energy functional E_{xc}:

$$F[\rho] = T[\rho] + \frac{1}{2}\iint drdr'\frac{\rho(r)\rho(r')}{|r - r'|} + E_{xc}[\rho] \qquad (2.28)$$

with the double integral equating to the classical Hartree energy E_H.

2.1.1.2 *Limitations*

While it is an amazing breakthrough for simplifying a many-electron problem, the HK theorems do not provide a systematic way of determining the form of some of the terms. The difficulties are in the kinetic energy and the electron-electron interaction. There is no explicit functional form for the kinetic energy. It is important to recognize that T can include correlations, such as backflow and the distorted motion of electrons in the presence of strong local moments due to f-electrons. In section 1.3.1.2, we approximated the electron-electron interaction V_{ee}, with the direct Coulomb and exchange interactions as given by the Hartree–Fock (HF) approximation. However, there are additional terms not accounted in the HF approximation. For example, the strong Coulomb interaction between two electrons with opposite spins exclude each other by a small volume. Such interactions are characterized as "non-local" in nature. An electron at r_1 produces a exclusion volume around r_1 called a Coulomb hole (COH). The pair correlation function $\rho_{\uparrow\downarrow}(r_1, r_2)$ describes the probability of finding an electron at r_2 when there is an electron with opposite spin moment at r_1. In a more general form, the pair correlation is given by

$$E_c^{\text{COH}} = \int dr_1 \rho_\sigma(r_1) \epsilon_{\uparrow\downarrow}(r_1) \tag{2.29}$$

$$\epsilon_{\uparrow\downarrow} = \int dr_2 \frac{\rho_{\uparrow\downarrow}(r_1, r_2)}{|r_2 - r_1|} \tag{2.30}$$

where ρ_σ is the σ-spin component of the electron density.

2.1.2 The Kohn–Sham equations

Kohn and Sham (1965) formulated a practical scheme for Eq. 2.22 in terms of a system of non-interacting particles $\{\phi_i(r)\}$ under the influence of an effective field, and further, provided a self-consistent algorithm for determining the ground state of electronic systems.

2.1.2.1 The single-particle picture

For non-interacting particles, the kinetic energy is the sum of individual particle kinetic energies of occupied states

$$T_0[\rho] = \sum_i^{\text{occ}} \int dr\, \phi_i^*(r) \left(-\frac{\hbar^2 \nabla^2}{2m} \right) \phi_i(r). \tag{2.31}$$

This single particle kinetic energy does not correctly account for the motion of the electrons due to XC, so they substituted the XC energy functional E_{xc} for the modified XC energy function E_{xc}'

$$E_{\text{xc}}'[\rho] = E_{\text{xc}}[\rho] + T[\rho] - T_0[\rho] \tag{2.32}$$

which subtracts the single particle kinetic energy and adds back the many electron kinetic energy. The new single-particle equations resulting from the variational principle are the Kohn–Sham (KS) equations:

$$\left[-\frac{\hbar^2}{2m} \nabla^2 + v_s(r) \right] \phi_i(r) = \epsilon_i \phi_i(r) \tag{2.33}$$

and

$$\rho(r) = \sum_i f_i \phi_i^*(r) \phi_i(r), \tag{2.34}$$

where

$$v_s(r) = v_{\text{ext}}(r) + \int dr' \frac{\rho(r')}{|r - r'|} + v_{\text{xc}}'[\rho(r)] \tag{2.35}$$

is the effective single-particle potential, or KS potential, f_i is the occupation number for state i and $v_{\text{xc}}' = \frac{\delta E_{\text{xc}}'}{\delta n}$.

2.1.2.2 *Kohn–Sham algorithm*

Once a suitable form of v'_{xc} is obtained, Eqs. 2.33 and 2.34 can be solved self-consistently by introducing a set of wave functions $\phi_i^{(\nu)}(\mathbf{r})$, eigenvalues $\epsilon_i^{(\nu)}$, densities $n^{(\nu)}(\mathbf{r})$, and potentials $v_s^{(\nu)}(\mathbf{r})$, characterizing the electron-electron interaction, that iteratively solve the KS equations

$$n^{(\nu)}(\mathbf{r}) = \sum_{i \in occ} \phi_i^{(\nu)*}(\mathbf{r})\phi_i^{(\nu)}(\mathbf{r}) \tag{2.36}$$

$$v_s^{(\nu)}(\mathbf{r}) = v_{ext}(\mathbf{r}) + \int \frac{n^{(\nu)}(\mathbf{r}')}{|\mathbf{r} - \mathbf{r}'|} d\mathbf{r}' + v'_{xc}[n^{(\nu)}(\mathbf{r})] \tag{2.37}$$

$$\epsilon_i^{(\nu+1)}\phi_i^{(\nu+1)}(\mathbf{r}) = \left[-\frac{\hbar^2}{2m}\nabla^2 + v_s^{(\nu)}(\mathbf{r}) \right] \phi_i^{(\nu+1)}(\mathbf{r}). \tag{2.38}$$

The procedure for determining the ground state density is as follows:

(1) Generate the initial density $\rho^{(0)}(\mathbf{r})$ from a guess, such as from the atomic densities.
(2) Use the electron density to calculate the Hartree potential and the effective XC potential to determine the effective single-particle potential $v_s^{(0)}(\mathbf{r})$.
(3) Solve Eq. 2.38 using the effective potential and determine the next set of wave functions $\phi_i^{(1)}(\mathbf{r})$ and eigenvalues $\epsilon_i^{(1)}$.
(4) Calculate the new density $n^{(1)}(\mathbf{r})$
(5) Go back to step (1) until the convergence criteria is met, such as the change in the total electronic energy $\left| E^{(\nu+1)} - E^{(\nu)} \right|$ is less than the desired numerical accuracy.

The solutions $\phi_i(\mathbf{r})$ represent the non-interacting, single-particle, orbitals that minimize the total energy.

2.1.3 *Approximate forms of the exchange-correlation functional*

The pair-correlation function $\varepsilon_{\uparrow\downarrow}$, introduced in Eq. 2.30, contains some of the necessary physics for determining a form of the exchange-correlation functional. The function $\rho_{\sigma\sigma'}(\mathbf{r}, \mathbf{r}')$ provides the probability density of finding an electron with spin σ at $\mathbf{r}$ and another electron with spin σ' at $\mathbf{r}'$. For the XC energy, a similar function exists as

$$E'_{xc}[\rho] = \frac{1}{2} \iint d\mathbf{r}d\mathbf{r}' \frac{\rho(\mathbf{r})\rho_{xc}(\mathbf{r}, \mathbf{r}')}{|\mathbf{r} - \mathbf{r}'|}. \tag{2.39}$$

The interpretation of ρ_{xc} is that if there is an electron at r', exchange statistics and correlation causes the classical density at r to reduce, creating a hole around r', or

$$\int dr' \rho_{xc}(r, r') = -1. \tag{2.40}$$

Eq. 2.40 is the sum-rule and is typically used to constrain approximations to ρ_{xc}.

2.1.3.1 *Local and non-local functionals*

In principle, the XC energy density is expressed as

$$\epsilon_{xc} = \int dr' \frac{\rho_{xc}(r, r')}{|r - r'|} \tag{2.41}$$

which involves a non-local pair correlation function. While a number of non-local pair-correlation functionals exist, (Gunnarsson *et al.*, 1976, for example), they are rarely used in practice. Instead, local and semi-local functionals are used which only rely on the density at r' or the derivative of the density at r', respectively. Three practical XC functionals are described in the following sections. They are the local-density approximation (LDA), the local-spin-density approximation (LSDA) and the generalized gradient approximation (GGA).

2.1.3.2 *The local density approximation*

For the LDA (Perdew and Wang, 1992; Cole and Perdew, 1982; Perdew and Zunger, 1981; Vosko *et al.*, 1980), ε'_{xc} is a simple scalar function of the density at r

$$\varepsilon'_{xc}[\rho(r)] = \varepsilon'_{xc}(\rho)|_{\rho=\rho(r)} \tag{2.42}$$

$$E'^{LDA}_{xc}[\rho] = \int dr \varepsilon'_{xc}[\rho(r)]\rho(r). \tag{2.43}$$

The energy density is determined, for example, by the values for the exchange and correlation derived or calculated using Monte Carlo (MC) methods on the homogeneous electron gas (HEG) (Ceperley and Alder, 1980). The effective XC potential is then

$$v'_{xc}[\rho(r)] = \frac{\delta E'^{LDA}_{xc}[\rho]}{\delta \rho(r)} \tag{2.44}$$

$$= \varepsilon'^{HEG}_{xc}[\rho(r)] + \rho(r) \left. \frac{\delta \varepsilon'^{HEG}_{xc}[\rho]}{\delta \rho} \right|_{\rho=\rho(r)}. \tag{2.45}$$

The XC energy density $\epsilon_{xc}[\rho]$ is a local function of r by fitting the MC result by Perdew and Zunger (1981). Introducing the WS radius $r_s(r)$ defined by

$$\frac{4\pi r_s^3}{3} = \frac{1}{\rho(r)}.$$

(2.46)

A practical form for the LDA energy density provided by Perdew and Zunger (1981) is

$$\epsilon_x[\rho(r)] = -\frac{0.4582}{r_s}$$

(2.47)

$$\epsilon_c[\rho(r)] = \begin{cases} -\dfrac{0.1423}{1 + 1.0529\sqrt{r_s} + 0.3334 r_s} & r_s \geq 1 \\ -0.048 + 0.311 \ln(r_s) - 0.0116 r_s + 0.002 r_s \ln(r_s) & r_s < 1. \end{cases}$$

(2.48)

2.1.3.3 *Local spin density approximation*

For a system with unequal spin contributions, such as the HMs and FMs discussed in chapter 1, the LSDA is employed to account for the separate spin channels. The basic idea of the LSDA is the same as the LDA; however, the expressions are complicated by the exchange. The generalized density for a two-spin system can be expressed as a density matrix:

$$\rho_{\alpha\beta}(r) = \sum_{j=\text{occ}} \phi_{\alpha,j}^*(r)\phi_{\beta,j}(r),$$

(2.49)

where the indices α and β denote either $\uparrow$ or $\downarrow$ with respect to a fixed z-axis. The total particle number is the trace of the density matrix, while the magnetic moment density m is the trace of the density matrix multiplied by the Pauli matrix

$$m(r) = \text{Tr}\left[\boldsymbol{\sigma}\rho(r)\right].$$

(2.50)

When the spin polarization is confined to the z-axis, such as the case of materials with weak or no SO coupling, only the diagonal elements of the density matrix, corresponding to the spin up density ($\rho_\uparrow = \rho_{\uparrow\uparrow}$) and the spin down density ($\rho_\downarrow = \rho_{\downarrow\downarrow}$), will be non-zero. The XC energy functional is then a functional of the two spin channels

$$E_{xc}^{\prime LSDA}[\rho_\uparrow, \rho_\downarrow] = \int dr \rho(r)\varepsilon_{xc}^{\prime LSDA}[\rho_\uparrow(r), \rho_\downarrow(r)]$$

(2.51)

$$= \int dr \rho(r)\varepsilon_{xc}^{\prime LSDA}[\rho(r), \zeta(r)]$$

(2.52)

where $\rho = \rho_\uparrow + \rho_\downarrow$ and ζ is the reduced spin polarization density

$$\zeta(\boldsymbol{r}) = \frac{\rho_\uparrow(\boldsymbol{r}) - \rho_\downarrow(\boldsymbol{r})}{\rho_\uparrow(\boldsymbol{r}) + \rho_\downarrow(\boldsymbol{r})}. \tag{2.53}$$

The KS equation must be modified to account for the spin index because the self-consistent part of the potential characterizes the interaction with the oppositely oriented spins differently and the energy must be minimized with respect to both spin densities simultaneously. The minimization of the energy gives

$$\sum_\nu \left\{ \left[-\frac{\hbar^2}{2m} \nabla^2 + v_{\text{ext}}(\boldsymbol{r}) + \sum_\gamma \int d\boldsymbol{r}' \frac{\rho_\gamma(\boldsymbol{r}')}{|\boldsymbol{r} - \boldsymbol{r}'|} \right] \delta_{\mu\nu} + v'_{\text{xc}}[\rho(\boldsymbol{r})]_\mu - \lambda \right\} \phi_\nu$$
$$= 0, \tag{2.54}$$

where the greek subscripts correspond to $\uparrow$ and $\downarrow$ and the effective XC potential is determined for each spin separately

$$v'_{\text{xc}}[\rho(\boldsymbol{r})]_\mu = \frac{\delta E'^{LSDA}_{\text{xc}}[\rho_\uparrow, \rho_\downarrow]}{\delta \rho_\mu(\boldsymbol{r})}. \tag{2.55}$$

For spin polarized densities, the magnetic moment density is simply the density difference in each spin channel

$$m(\boldsymbol{r}) = \rho_\uparrow(\boldsymbol{r}) - \rho_\downarrow(\boldsymbol{r}), \tag{2.56}$$

and the magnetic moment is

$$M = \int d\boldsymbol{r} \, m(\boldsymbol{r}). \tag{2.57}$$

2.1.3.4 *The generalized gradient approximation*

The LDA underestimates the lattice constant and the gap for semiconductors and insulators. To improve upon the LDA, the next level of XC approximations include the GGA, first developed by Langreth and Perdew (1980) to include the contribution of the derivative of ρ with respect to $\boldsymbol{r}$. The GGA accounts for the slope of ρ at $\boldsymbol{r}$

$$\epsilon_{xc}[\rho(\boldsymbol{r}), \nabla \rho(\boldsymbol{r})]. \tag{2.58}$$

This form is still local in that it depends only on the coordinate $\boldsymbol{r}$, however some call it semilocal (Perdew *et al.*, 1992). These approximations have the spin polarized form

$$E'^{GGA}_{\text{xc}}[\rho] = \int d\boldsymbol{r} \, \varepsilon_{\text{xc}}(\rho_\uparrow, \rho_\downarrow, |\nabla \rho_\uparrow|, |\nabla \rho_\downarrow|) \rho(\boldsymbol{r}). \tag{2.59}$$

GGA calculations tend to improve total energies, structural energy differences and sometimes the bonding energies over the LDA (Perdew *et al.*, 1996). Additionally, the GGA tends to overestimate the lattice constants of semiconductors (Filippi *et al.*, 1994).

Among the different forms of the GGA, the Perdew–Burke–Ernzerhof (PBE) (Perdew, Burke and Ernzerhof, 1996) GGA is one of the most popular and has the form

$$E_{\mathrm{xc}}[\rho] = E_c + E_x \tag{2.60}$$

$$E_c[\rho_\uparrow, \rho_\downarrow] = \int d\mathbf{r}\rho(\mathbf{r}) \left[\varepsilon_c^{\mathrm{HEG}}(r_s, \zeta) + H_c(r_s, \zeta, t)\right] \tag{2.61}$$

$$E_x[\rho_\uparrow, \rho_\downarrow] = \int d\mathbf{r}\rho(\mathbf{r}) \left[\varepsilon_x^{\mathrm{HEG}}(\rho) F_x(s)\right] \tag{2.62}$$

where r_s is the WS radius, ζ is the reduced polarization, t and s are reduced gradient terms ($t = |\nabla\rho|/2k_s\rho$ and $s = |\nabla\rho|/2k_F\rho$, where k_s and k_F are the WS and Fermi $\mathbf{k}$-vectors, respectively). The benefit of the PBE form of GGA is that the functions H_c and F_x are simple analytic functions that correctly reproduce a number of limiting cases and scaling rules (Perdew *et al.*, 1996, and references therein). The analytic forms of H_c and F_x are detailed below.

For H_c in Eq. 2.61, Perdew *et al.* used the form

$$H_c = \frac{e^2}{a_0}\gamma\phi^3 \ln\left\{1 + \frac{\beta}{\gamma}t^2\left[\frac{1 + At^2}{1 + At^2 + A^2t^4}\right]\right\}, \tag{2.63}$$

$$A = \frac{\beta}{\gamma}\left[\exp\left(-\varepsilon_C^{\mathrm{HEG}}/\gamma\phi^3 e^2/a_0\right) - 1\right]^{-1}, \tag{2.64}$$

where $\beta \approx 0.066725$, $\gamma \approx 0.031091$ and ϕ is the spin-scaling factor

$$\phi = \left(\frac{\varepsilon_c^{\mathrm{HEG}}(\zeta)}{\varepsilon_c^{\mathrm{HEG}}(\zeta = 0)}\right)^{1/3} \tag{2.65}$$

$$\approx \left[(1 + \zeta)^{2/3} + (1 - \zeta)^{2/3}\right]/2 \tag{2.66}$$

approximated by Wang and Perdew. H_c is designed to satisfy three conditions:

- The slowly varying density limit $t \to 0$: This limit should recover the LSDA when $t = 0$ ($H_c = 0$) (Wang and Perdew, 1991).
- The rapidly varying density limit $t \to \infty$: In this limit, the correlation energy tends to zero rapidly. This is due to the cancellation of contributions to the correlation energies at very high kinetic energies.

- The high density limit $\rho \to \infty$: Correlation is mainly due to screening effects in this limit and H_c reproduces the correct screening behavior for large and slowly varying densities (Ma and Brueckner, 1968).

Wang and Perdew also suggested

$$F_x(s) = 1 + \kappa - \frac{\kappa}{1 + \mu s^2/\kappa} \tag{2.67}$$

with $\kappa \approx 0.804$ and $\mu \approx 0.21951$. This expression also satisfies three conditions:

- The spin-scaling relationships for polarized $E_x[\rho_\uparrow, \rho_\downarrow]$ and unpolarized $E_x[\rho]$ functionals: The unpolarized exchange energy should be the sum of the polarized exchange energy with half the electrons in one spin channel and half in the other spin channel

$$E_x[\rho] = E_x[\rho/2, 0] + E_x[0, \rho/2] \tag{2.68}$$

$$E_x[\rho_\uparrow, \rho_\downarrow] = E_x[\rho_\uparrow, 0] + E_x[0, \rho_\downarrow] \tag{2.69}$$

$$= \frac{1}{2}\left(E_x[2\rho_\uparrow] + E_x[2\rho_\downarrow]\right). \tag{2.70}$$

- The expression reproduces the LSDA exchange energy at nearly uniform density. Investigations into gradient corrections to LSDA by Ortiz (1992) demonstrate that the LSDA by itself gives better agreement with many-body calculations. Ortiz found that the gradient corrections incorrectly introduce a repulsive term that produces incorrect screening behavior in the nearly uniform density limit.
- The Lieb-Oxford bound (Lieb and Oxford, 1981): The Lieb-Oxford bound takes the form

$$E \geq -Ce^{2/3} \int \rho^{4/3}(\boldsymbol{r}) d\boldsymbol{r}, \tag{2.71}$$

where $C = 1.679$ was the provable lower limit on the Coulomb energy at the time of the publication of PBE (1991). The limit follows from a plane wave calculation of spinless fermions in a box (Lieb, 1979).

2.1.4   *The augmented plane wave method*

After determining the form of the electron-electron interaction by implementing some form of the XC potential above, the external potential v_{ext} needs to be specified to determine the electronic and magnetic properties of a physical system.

An atom placed in a crystal has two general types of electrons: tightly bound, lower energy, core electrons and spread out, higher energy, valence electrons. The valence electrons have very high oscillations near the core because they must be orthogonal to the core states but they spread out in the interstitial regions between atoms. Additionally, the tightly bound core states in a crystal play essentially no role in the electronic properties of a crystal and tend to change little when compared to the same states in the atom.

While the core states are generally known from atomic calculations, the valence states present a problem: due to the vastly different behavior of the valence states near and away from the core, is there an efficient way to compute these states? Due to the large oscillations near the core, plane waves are ineffective for describing valance states as many plane waves would be necessary to describe the oscillations. Additionally, localized states are ineffective at describing the spread out behavior. Two popular methods have been developed to address this issue. The first is the augmented plane wave (APW) method and the second is the pseudopotential method. The APW method will be discussed first.

2.1.4.1 *Augmented plane waves*

In 1937, Slater introduced the muffin-tin potential approximation for the ionic potential and the APW. This approximation splits the ionic potential into two regions. Near the ion, within a sphere called the muffin-tin (MT) sphere, he used the ionic potential while outside the MT sphere, he assumed the potential to be a constant since the potential varies much less rapidly than near the ion. The MT spheres for adjacent ions do not overlap. An example of the MT potential in 2 dimensions is shown in Fig. 2.1

Similarly, he introduced an expansion of the single-particle valence wave function in terms of two sets of basis functions. One set, comprised of plane waves, spans the interstitial region between atoms (outside the MT sphere) while the other set augment the core region near each atom (inside the MT sphere). An augmented plane wave has the form

$$\psi_{\boldsymbol{k}+\boldsymbol{G}_n}^{\text{APW}}(\boldsymbol{r}) = \begin{cases} e^{i(\boldsymbol{k}+\boldsymbol{G}_n)\cdot\boldsymbol{r}} & r > R_{\text{MT}} \\ \sum_{lm} C_{lm}(\boldsymbol{k}+\boldsymbol{G}_n)i^l Y_{lm}(\hat{\boldsymbol{r}})\psi_{lm}(\epsilon, r) & r < R_{\text{MT}} \end{cases} \quad (2.72)$$

where ψ_{lm} is a solution to the radial Schrödinger equation with band energy ϵ, $\boldsymbol{k}$ are wave numbers within the first BZ, $\boldsymbol{G}_n$ are RLVs, and C_{lm} coefficients are determined by the boundary conditions at the MT sphere. Slater

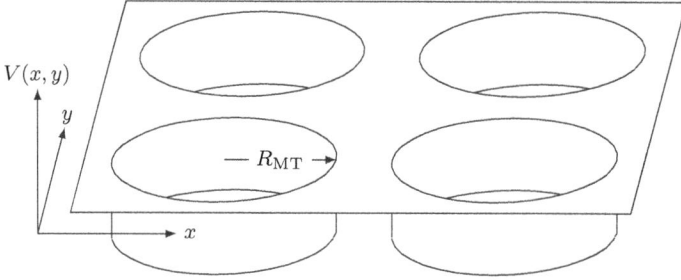

Fig. 2.1   Portion of the muffin-tin potential in 2 dimensions. Outside this radius, the potential is zero. In general, the potential inside the muffin-tin radius can be the ionic potential.

(1937) recognized that the APW have discontinuous slopes at $r = R_{MT}$ so he needed to treat the kinetic energy carefully. Commonly, the Laplacian $\nabla^2 \Psi$ is used in the kinetic energy, in which case the kinetic energy of the APWs is undefined at the MT radius. Instead, Slater used the more fundamental relation $\nabla \Psi \cdot \nabla \Psi$ which avoids the problem of the discontinuity of the slope.

2.1.4.2   *Issues with the APW method*

Koelling (1975) identified two issues arising from the use of the APW functions. The first issue is the appearance of the energy ϵ in the wave function resulting in off-diagonal matrix elements of the Hamiltonian that were also dependent on the energy. Special care must be taken when solving for the roots of the secular equation since it is no longer linear. The second issue occurs for special values of ϵ where a node of the radial solution coincides with the radius of the MT sphere. Near this energy, the plane wave part cannot easily match the boundary condition and the solution varies rapidly. The second issue is called the "asymptote problem".

The first issue with APW method, that the energy ϵ appears in every matrix element of the Hamiltonian, has a variety of computationally expensive solutions (Koelling, 1972). One solution, given by Koelling (1972), is that the basis functions are slowly varying functions of the energy. The logarithmic derivative at R_{MT} could then be linearly approximated by $a_l - b_l(\epsilon - \epsilon_0)$, where ϵ_0 is used to select the energy region of interest. The coefficients a_l and b_l are determined by fitting the logarithmic derivative at the MT boundary. This solution does not address the second issue.

2.1.4.3 *The linearized augmented plane wave method*

A more robust solution to the issues presented in the APW method was developed by Koelling (1975) and Andersen (1975). They solved both issues by combining the radial solution with its energy derivative inside the MT sphere for a particular l-value. The new test solutions are

$$\psi(\mathbf{k}) = \sum_{lm} \left[A_{lm} u_l(\epsilon_l) + B_{lm} \dot{u}_l(\epsilon_l) \right] Y_{lm}(\hat{r}) \tag{2.73}$$

inside the MT spheres. The advantage of using the energy derivative is that, as shown later, the radial solutions and energy derivatives are orthogonal. Let the secular equation inside the MT sphere be

$$h_l u_l(\epsilon_l, r) = \left[-\frac{1}{r^2} \frac{\partial}{\partial r} \left(r^2 \frac{\partial}{\partial r} \right) + \frac{l(l+1)}{r^2} + v(r) \right] u_l(\epsilon_l, r) \tag{2.74}$$

$$= \epsilon_l u_l(\epsilon_l, r). \tag{2.75}$$

The normalization condition is

$$\int_0^{R_{\mathrm{MT}}} r^2 dr u_l^2(\epsilon_l, r) = 1. \tag{2.76}$$

If the radial eigenvalue is treated as a parameter $\epsilon_l \to \epsilon$ and Eq. 2.76 is differentiated with respect to ϵ, one can show that $\dot{u}_l = \frac{\partial u_l}{\partial \epsilon}$ and u_l are mutually orthogonal:

$$\langle u_l \,|\, \dot{u}_l \rangle = \langle \dot{u}_l \,|\, u_l \rangle = 0 \tag{2.77}$$

Differentiating Eq. 2.75 with respect to ϵ gives differential equation for the energy derivative of u_l

$$(h_l - \epsilon)\, \dot{u}_l = u_l. \tag{2.78}$$

The functions u_l and $\dot{u}_l$ do not span the same Hilbert space, so the matrix elements

$$\langle u_l \,|\, h_l - \epsilon \,|\, \dot{u}_l \rangle = 1 \tag{2.79}$$

$$\langle \dot{u}_l \,|\, h_l - \epsilon \,|\, u_l \rangle = 0 \tag{2.80}$$

are not Hermitian. $\dot{u}_l$ may be obtained from Eq. 2.78 to within a normalization constant since

$$\int_0^{R_{\mathrm{MT}}} r^2 dr \dot{u}_l^2 = N_l \tag{2.81}$$

is not necessarily unity. However the normalization factor is found using the solutions to the matrix elements and subtracting Eqs. 2.79 and 2.80

$$\langle u_l \,|\, h_l \,|\, \dot{u}_l \rangle - \langle \dot{u}_l \,|\, h_l \,|\, u_l \rangle = 1 \tag{2.82}$$

and Green's theorem (integration by parts) within the MT sphere. The normalization condition is

$$R_{\mathrm{MT}}^2 \left[u_l'(R_{\mathrm{MT}})\dot{u}_l(R_{\mathrm{MT}}) - u_l(R_{\mathrm{MT}})\dot{u}_l'(R_{\mathrm{MT}}) \right] = 1 \qquad (2.83)$$

where the prime denotes differentiation of the radial coordinate r. Using the requirement that each angular momentum l function is continuous with continuous derivatives across the MT boundary, the solutions in Eq. 2.73 have coefficients

$$A_{lm}(\boldsymbol{k}) = \frac{4\pi R_{\mathrm{MT}}^2}{\sqrt{V}} i^l Y_{lm}^*(\hat{\boldsymbol{k}}) \left[j_l'(kR_{\mathrm{MT}})\dot{u}_l - j_l(kR_{\mathrm{MT}})\dot{u}_l' \right] \qquad (2.84)$$

$$B_{lm}(\boldsymbol{k}) = \frac{4\pi R_{\mathrm{MT}}^2}{\sqrt{V}} i^l Y_{lm}^*(\hat{\boldsymbol{k}}) \left[j_l(kR_{\mathrm{MT}})u_l' - j_l'(kR_{\mathrm{MT}})u_l \right]. \qquad (2.85)$$

Equations 2.73–2.85 resolve the second issue with the APW method, the asymptote problem, by never letting $\psi(\boldsymbol{k})$ be zero at the MT sphere boundary. The orthogonality condition (Eq. 2.77) prevents $u_l(R_{\mathrm{MT}})$ and $\dot{u}_l(R_{\mathrm{MT}})$ from being zero simultaneously. Since the radial function never has a node at the boundary, it is then possible to match the function with the plane wave at the boundary.

2.1.5   *The pseudopotential method*

The main goal of the pseudopotential method is to replace the effects of core states with a repulsive and energy dependent potential $\hat{V}_r$ that a valence electron feels. The repulsive potential smooths out the valence electron $|\phi_v\rangle$ oscillations near the core. The method has two benefits for numerical calculations: first, the core electrons are removed, so there are fewer states to store, and second, the new valence states $|\tilde{\phi}_v\rangle$, called the pseudovalence states, are smoother so they may be treated more efficiently using plane waves. For the new states, the Hamiltonian for a pseudopotential model is

$$H_p = -\frac{1}{2}\nabla^2 + U + \hat{V}_r \qquad (2.86)$$

$$= -\frac{1}{2}\nabla^2 + \hat{V}_p \qquad (2.87)$$

where U is the usual potential energy in the KS formulas, $\hat{V}_r$ is a repulsive potential in the core region that weakens the overall potential, and $\hat{V}_p$ is the pseudopotential. $\hat{V}_r$ is energy-dependent and non-local.

2.1.5.1 *Orthogonalized plane waves*

Herring (1940) developed the orthogonalized plane wave (OPW) method as a practical way to treat the effects of the core electrons. In his description, Herring constructed a OPW, to be like plane waves, but that are orthogonal to the localized core states

$$|\phi_{\mathrm{OPW}}\rangle = |\boldsymbol{k}\rangle - \sum_c |\phi_c\rangle \langle \phi_c \,|\, \boldsymbol{k}\rangle \tag{2.88}$$

$$= (1 - P_c) \,|\boldsymbol{k}\rangle \tag{2.89}$$

where $|\boldsymbol{k}\rangle$ is a plane wave, $|\phi_c\rangle$ denotes an atomic core state obeying the Hamiltonian equation $H \,|\phi_c\rangle = E_c \,|\phi_c\rangle$ and P_c is the projection operator of the core states. The orthogonalization procedure becomes tedious with large atoms with many core states.

2.1.5.2 *Pseudopotentials*

Phillips and Kleinman (1959) reformulated the OPW method by replacing the plane wave with a smooth wave function called the pseudowave function $|\tilde{\phi}\rangle$. The wave function for an electron is then

$$|\phi\rangle = \left|\tilde{\phi}\right\rangle - \sum_c |\phi_c\rangle \left\langle \phi_c \,\middle|\, \tilde{\phi} \right\rangle \tag{2.90}$$

$$\tag{2.91}$$

Schrödinger's equation for $|\phi\rangle$ is

$$\hat{H} \,|\phi\rangle = \hat{H} \left|\tilde{\phi}\right\rangle - \sum_c E_c \,|\phi_c\rangle \left\langle \phi_c \,\middle|\, \tilde{\phi} \right\rangle \tag{2.92}$$

$$= \left(\hat{T} + \hat{V} - \sum_c E_c \,|\phi_c\rangle \langle \phi_c| \right) \left|\tilde{\phi}\right\rangle \tag{2.93}$$

$$= E \left|\tilde{\phi}\right\rangle - E \sum_c |\phi_c\rangle \left\langle \phi_c \,\middle|\, \tilde{\phi} \right\rangle. \tag{2.94}$$

Rearranging this expression produces

$$\hat{H} \left|\tilde{\phi}\right\rangle + \sum_c (E - E_c) \,|\phi_c\rangle \left\langle \phi_c \,\middle|\, \tilde{\phi} \right\rangle = E \left|\tilde{\phi}\right\rangle \tag{2.95}$$

$$\left(\hat{T} + \hat{V}_{psp}\right) \left|\tilde{\phi}\right\rangle = E \left|\tilde{\phi}\right\rangle \tag{2.96}$$

where

$$\hat{V}_{psp} = \hat{V} + \sum_c (E - E_c) \,|\phi_c\rangle \langle \phi_c| \tag{2.97}$$

is the pseudopotential. This potential acts to weaken the overall ionic Coulomb potential in the core region thereby producing much smoother functions $|\tilde{\phi}\rangle$ that are easily expanded on a plane wave basis.

2.1.5.3 *Nonuniqueness of the pseudopotential*

The general psuedopotential theorem, as developed by Austin *et al.* (1962), demonstrates the non-uniqueness of the pseudopotential.

Theorem 2.3 (General pseudopotential theorem).
The pseudo-Hamiltonian H_p has the same valence eigenvalues as the real Hamiltonian H.

Proof. The psuedowave and real wave functions obey Schrödinger's equation

$$(H - E_n)|\phi_n\rangle = 0 \tag{2.98}$$

$$(H + \hat{V}_{psp} - \tilde{E}_n)|\tilde{\phi}_n\rangle = 0, \tag{2.99}$$

where E_n are the eigenvalues of the full Hamiltonian, $\tilde{E}_n$ represents the eigenvalues of the pseudowave states and n can represent either core c or valence v states. Following the original proof by Austin *et al.*, we assume that the real wave functions are complete and non-degenerate and define the general pseudopotential

$$\hat{V}_{psp}|\tilde{\phi}_n\rangle = \sum_c |\phi_c\rangle \left\langle F_c \middle| \tilde{\phi}_n \right\rangle \tag{2.100}$$

where $|F_c\rangle$ are arbitrary functions that, in general, give $\langle F_c|\tilde{\phi}_n\rangle \neq 0$. We expand the pseudowave states as linear combinations of real wave functions

$$|\tilde{\phi}_n\rangle = \sum_{n'} a_{nn'}|\phi_{n'}\rangle \tag{2.101}$$

and substitute them into Eqs. 2.100 and 2.99. Schrödinger's equation becomes

$$\sum_{n'} a_{nn'}\left[\left(E_{n'} - \tilde{E}_n\right)|\phi_{n'}\rangle + \sum_{c''}|\phi_{c''}\rangle \langle F_{c''}|\phi_{n'}\rangle\right] = 0. \tag{2.102}$$

We first consider the case $n = c$. Eq. 2.102 becomes

$$\sum_{c'} a_{cc'}\left[\left(E_{c'} - \tilde{E}_c\right) + \sum_{c''}|\phi_{c''}\rangle \langle F_{c''}|\right]|\phi_{c'}\rangle$$
$$+ \sum_{v'} a_{cv'} \sum_{c''}\langle F_{c''}|\phi_{v'}\rangle|\phi_{c''}\rangle$$
$$+ \sum_{v'} a_{cv'}\left(E_{v'} - \tilde{E}_c\right)|\phi_{v'}\rangle = 0, \tag{2.103}$$

and therefore $a_{cv'}$ are all zero except in the unlikely situation where both $\sum_{v'} \langle F_{c''} | \phi_{v'} \rangle$ and $E_{v'} - \tilde{E}_c$ are simultaneously zero. The pseudocore state only depends on the real core wave functions

$$|\tilde{\phi}_c\rangle = \sum_{c'} a_{cc'} |\phi_{c'}\rangle \tag{2.104}$$

Similarly, when we let $n = v$ in Eq. 2.102, we find the terms

$$a_{vv}(E_v - \tilde{E}_v) |\psi_v\rangle + \sum_{v' \neq v} a_{vv'}(E_{v'} - \tilde{E}_v) |\phi_{v'}\rangle + (\text{core state terms}) = 0. \tag{2.105}$$

The first term shows $E_v = \tilde{E}_v$—the condition that we set out to prove. Additionally, without degeneracy, the second term implies that $a_{vv'} = 0$. The pseudovalence state is related to the real valance plus a linear combination of the real core states

$$|\tilde{\phi}_v\rangle = |\phi_v\rangle + \sum_{c'} a_{vc'} |\phi_{c'}\rangle. \tag{2.106}$$

$\square$

The general pseudopotential theorem is extremely important for numerical calculations. The choice of the pseudocore functions does not have an impact on the valence eigenvalues so these functions and the pseudopotential can be chosen to reduce the overall computational effort.

2.1.5.4 Nonlocal pseudopotential

In the semi-nonlocal form of the pseudopotential, the non-local projection operator only involve spherical harmonics Y_{lm}

$$V_R = V(\mathbf{r}) + \sum_{lm} |Y_{lm}\rangle V_l(r) \langle Y_{lm}| \tag{2.107}$$

as used by Lee and Falicov (1968) to calculate the band structure of potassium. A $\mathbf{k}$-dependent form was used by Fong and Cohen (1970) to calculate the band structure of copper. The radial part of the semi-local projection is not separable, however Kleinman and Bylander (1982) proposed an approximate separable form of the non-local operators to reduce the computation and storage costs. In their approximation, non-separable integrals of the form

$$\langle \mathbf{k}' | Y_{lm} \rangle V_l(r) \langle Y_{lm} | \mathbf{k} \rangle \sim \int r^2 dr \, j_l(kr) j_l(k'r) V_l(r) P_l(\cos \theta_{\mathbf{k}\mathbf{k}'}) \tag{2.108}$$

are replaced with the separated integrals

$$\langle \mathbf{k}' \,|\, v_l \rangle \langle v_l \,|\, \mathbf{k} \rangle \sim \int r^2 \, dr j_l(kr) v_l(r) \int r^2 \, dr j_l(k'r) v_l(r) P_l(\cos \theta_{\mathbf{k}\mathbf{k}'}) \quad (2.109)$$

where j_l are the spherical Bessel functions and P_l are the Legendre polynomials. An approximate form of $v_l(r)$ is

$$v_l(r) \approx V_l(r) \tilde{\phi}_l(r) / \sqrt{\left\langle \tilde{\phi}_l \,\middle|\, V_l(r) \,\middle|\, \tilde{\phi}_l \right\rangle} \quad (2.110)$$

where $\tilde{\phi}_l(r)$ is the pseudovalence wavefunction.

2.1.5.5 *Norm-conserving psuedopotentials*

Outside the core region the pseudo wave functions are not necessarily normalized to match the real wave functions. The amount of charge inside the core region is not correct and the l-components of the pseudowave function logarithmic derivatives do not match the real ones. This makes pseudopotentials difficult to transfer to new crystal and atomic environments. Norm-conserving pseudopotentials, introduced by Hamann *et al.* (1979), address this shortcoming by considering that the pseudopotential should have the property that the pseudowave functions are identical to the real wave functions (as well as the logarithmic derivatives) everywhere outside the core radius and that the norm of the pseudowave and real wave functions are identical

$$\left\langle \tilde{\phi}_i \,\middle|\, \tilde{\phi}_j \right\rangle = \langle \phi_i \,|\, \phi_j \rangle . \quad (2.111)$$

Inside the core region, the exact form of the pseudopotential does not matter as long as the amount of charges the valence wave functions "sees" remains the same and the logarithmic derivatives match.

2.1.5.6 *Ultra-soft pseudopotentials*

For some pseudopotentials, the pseudowave function, constructed using the norm-conserving condition, could not be smoother than the real wave functions and did not greatly improve the computational cost. These pseudopotentials are necessarily very strong and generate large oscillations in their respective pseudowave functions that are not easily constructed by a few plane waves. To address this issue, Vanderbilt (1990) developed ultra-soft (US) pseudopotentials that were both transferable to different crystal environments and were much smoother than the norm-conserving pseudopotentials. The norm-conserving scheme is unnecessary with US pseudopotential and the core radius could be extended well beyond the largest oscillation peak as shown in Fig. 2.2. Vanderbilt (1990) introduced an overlap term

$$\left\langle \tilde{\phi}_i \,\middle|\, 1 + Q_{ij} \,\middle|\, \tilde{\phi}_j \right\rangle = \left\langle \phi_i \,\middle|\, \phi_j \right\rangle \tag{2.112}$$

that defines the US pseudowave functions. This transformation removes a large contribution to the density that must be reintroduced for the calculation of the valence density. Later, Blöchl (1994) developed the projector augmented wave (PAW) method which is a generalization of the US method.

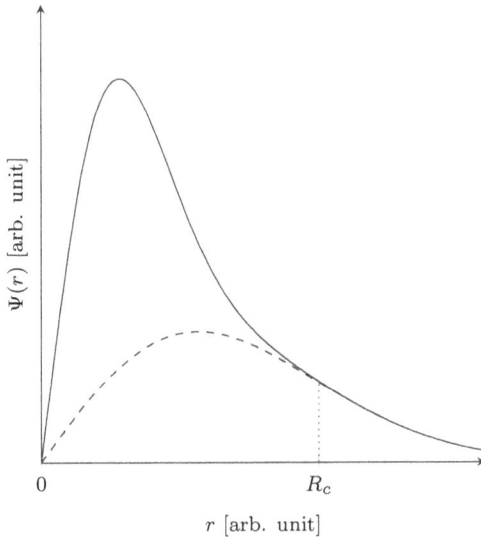

Fig. 2.2 Schematic of radial pseudowave functions from norm-conserving (solid-line) and ultrasoft (dashed-line) pseudopotentials. Above the cutoff radius R_c, the two functions are identical while below R_c, the ultrasoft function is much smoother. The pseudowave function constructed from the ultrasoft pseudopotential has a smaller density in the core region compared to the norm-conserving one.

2.1.5.7 *Projector augmented wave method*

The PAW method combines the ideas of pseudopotentials and linearized augmented plane wave (LAPW). Blöchl (1994) considered a linear transformation $\mathcal{T}$ from all-electrons (AEs) single particle wave functions ψ to pseudowave functions $\tilde{\psi}$:

$$|\psi\rangle = \mathcal{T} |\tilde{\psi}\rangle \tag{2.113}$$

where he assumed the form

$$\mathcal{T} = 1 + \sum_i \tilde{\mathcal{T}}_i \tag{2.114}$$

so the pseudowave functions differ from the AE ones by a linear transformation around the ion cores. The transformations $\mathcal{T}_i$ alter the wave function around the core at $\boldsymbol{R}_i$, the position of the ion, up to a radius R_c so the pseudo and AE wave functions are identical outside R_c. Within R_c, the augmentation region, the pseudowave function $|\tilde{\psi}\rangle$ and full wave function $|\psi\rangle$ can be expanded into a linear combination of pseudo valance states $|\tilde{\phi}_n\rangle$ and valence states $|\phi_n\rangle$, respectively, centered around $\boldsymbol{R}_i$

$$|\tilde{\psi}\rangle = \sum_n c_n |\tilde{\phi}_n\rangle \tag{2.115}$$

$$|\psi\rangle = \sum_n c_n |\phi_n\rangle \tag{2.116}$$

$$|\psi\rangle = |\tilde{\psi}\rangle + \sum_n c_n \left[|\phi_n\rangle - |\tilde{\phi}_n\rangle \right]. \tag{2.117}$$

Next, Blöchl introduced the projection operators $|p_n\rangle$ that give the coefficients of the linear expansion

$$\langle p_n | \tilde{\psi}\rangle = c_n. \tag{2.118}$$

These projection operators have the form

$$|p_n\rangle = \sum_m \left[\langle f_j | \tilde{\phi}_k \rangle \right]^{-1}_{nm} |f_m\rangle \tag{2.119}$$

where $|f_m\rangle$ are arbitrary, linearly independent functions and $[\langle f_j | \tilde{\phi}_k \rangle]^{-1}_{nm}$ denotes the matrix inverse of $\langle f_j | \tilde{\phi}_k \rangle$. The projection operator gives the form of the linear transformation $\mathcal{T}$

$$|\psi\rangle = |\tilde{\psi}\rangle + \sum_n \left[|\phi_n\rangle \langle p_n | \tilde{\psi}\rangle - |\tilde{\phi}_n\rangle \langle p_n | \tilde{\psi}\rangle \right] \tag{2.120}$$

$$= \left(1 + \sum_n \mathcal{T}_n \right) |\tilde{\psi}\rangle \tag{2.121}$$

$$\mathcal{T}_n = \left(|\phi_n\rangle - |\tilde{\phi}_n\rangle \right) \langle p_n|. \tag{2.122}$$

The main advantage of the PAW method is that expectation values of the AE system may be determined using the pseudowave functions and the transformation $\mathcal{T}$

$$\langle \phi | A | \phi \rangle = \langle \tilde{\phi} | \mathcal{T}^\dagger A \mathcal{T} | \tilde{\phi} \rangle \tag{2.123}$$

$$= \langle \tilde{\phi} | \tilde{A} | \tilde{\phi} \rangle \tag{2.124}$$

where

$$\tilde{A} = \mathcal{T}^\dagger A \mathcal{T} \tag{2.125}$$

$$= A + \sum_{\langle ij \rangle} |p_i\rangle \left(\langle \phi_i|A|\phi_j\rangle - \langle \tilde{\phi}_i|A|\tilde{\phi}_j\rangle \right) \langle p_j| \tag{2.126}$$

for local operators A (such as the density ρ), and the sum over $\langle ij \rangle$ denotes that the sum is taken over terms in the same augmentation region. The density ρ, for example, is separated into three parts

$$\langle \phi|\rho|\phi \rangle = \langle \tilde{\phi}|\rho|\tilde{\phi} \rangle + \sum_{\langle ij \rangle} \rho_{ij} \langle \phi_i|\rho|\phi_j \rangle - \sum_{\langle ij \rangle} \rho_{ij} \langle \tilde{\phi}_i|\rho|\tilde{\phi}_j \rangle \tag{2.127}$$

where

$$\rho_{ij} = \sum_n \langle \tilde{\phi}_n|p_i \rangle \langle p_j|\tilde{\phi}_n \rangle \tag{2.128}$$

is the occupancies of the augmentation regions (Kresse and Joubert, 1999). The first term is just the density of the pseudowave functions. The other two terms denote the charge densities of the AE and pseudowave functions within the augmentation regions, respectively.

2.1.6 *Linear response theory*

Methods based on DFT provide ways to determine ground state properties. The KS orbitals, however, bear little physical significance, in particular when trying to determine excited state properties (Perdew *et al.*, 1982). The fundamental gap, for example, is underestimated using LDA by at least 10%. Silicon has a band gap at the Γ-point of 3.4 eV (Welkowsky and Braunstein, 1972); the LDA results in a value around 2.6 eV (Hybertsen and Louie, 1985). Also the indirect gap of Si is 1.17 eV while the LDA gives 0.52 eV. Moreover, the GGA functional does not significantly improve the situation. Why are the excited state properties so poorly described by XC approximations? The main shortcoming, described by Perdew *et al.* (1982), is the failure of the XC functionals to correctly account for fractional occupation of the KS orbitals. Many modern techniques exist that correct the band gap problem caused by the LDA and GGA approximations. Some of these techniques involve adjustable parameters that are fixed relative to known experimental results. Hybrid functionals blend together multiple XC functionals by an amount that provides reasonable agreement with experiments. The PBE0 hybrid functional (Adamo and Barone, 1999), for example, mixes 25% exact exchange (Görling, 1996), 75% exchange from the PBE (Perdew *et al.*, 1996) functional and 100% correlation from PBE.

Instead of relying on adjustable parameters, a few *ab initio* methods exist based on linear response (LR) theory and Green's function methods. The so-called *GW* method, developed by Hedin (1965), is a very popular scheme for determining band widths and the fundamental band gap that give results in close agreement with experimental values. The theory relies on LR introduced in the following sections.

2.1.6.1 *Linear susceptibility*

The linear response of a system describes how the system responds to weak external fields. In the linear response regime, the magnitude of the response is directly proportional to the magnitude of the external field. The non-local linear response is characterized by the response function χ

$$\langle \delta A(\boldsymbol{r}, t) \rangle = \int d\boldsymbol{r}' dt' \, \chi(\boldsymbol{r}, \boldsymbol{r}'; t, t') f(\boldsymbol{r}', t') \tag{2.129}$$

where f is related to the field strength and the δA denotes the fluctuation away from the ground state expectation value. As a specific example used throughout the remainder of these notes, the response function χ characterizes the fluctuation in electron density $\delta\rho$ due to an externally applied electric potential v_{ext}

$$\langle \delta\rho(\boldsymbol{r}, t) \rangle = \int d\boldsymbol{r}' dt' \, \chi(\boldsymbol{r}, \boldsymbol{r}'; t, t') v_{\text{ext}}(\boldsymbol{r}', t'). \tag{2.130}$$

The variation in the charge density is the polarization of the material, which induces a local, microscopic, electric field. The overall effective field is the sum of external and induced fields

$$v_{\text{eff}}(\boldsymbol{r}, t) = v_{\text{ext}}(\boldsymbol{r}, t) + \int d\boldsymbol{r}' v(|\boldsymbol{r} - \boldsymbol{r}'|) \langle \delta\rho(\boldsymbol{r}', t') \rangle \tag{2.131}$$

where v is the Coulomb potential. A similar property to χ is the inverse dielectric function ϵ^{-1} which describes the effective potential due to screening of an arbitrary v_{ext},

$$v_{\text{eff}}(\boldsymbol{r}, t) = \int d\boldsymbol{r}' dt' \epsilon^{-1}(\boldsymbol{r}, \boldsymbol{r}'; t, t') v_{\text{ext}}(\boldsymbol{r}', t') \tag{2.132}$$

$$\epsilon^{-1}(\boldsymbol{r}, \boldsymbol{r}'; t, t') = \delta(\boldsymbol{r} - \boldsymbol{r}')\delta(t - t') + v(|\boldsymbol{r} - \boldsymbol{r}'|)\chi(\boldsymbol{r}, \boldsymbol{r}'; t, t') \tag{2.133}$$

These functions are useful in excited state calculations because they accurately accounts for how electrons screen the long-ranged Coulomb interaction.

2.1.6.2 *Coulomb screening*

When discussing many-body physics, it is sometimes customary to use the following standard shorthand notation for the coordinates, integrals and functions:

$$f(12) = f(\boldsymbol{r}_1, t_1, \boldsymbol{r}_2, t_2)$$

$$\int d1 = \int d\boldsymbol{r}_1 dt_1$$

$$f(1^+) = f(\boldsymbol{r}, t + \eta)$$

where η is an infinitesimal positive number.

The premise behind screening is that when a negative point charge is introduced to a collection of electrons, the electrons rearrange in the presence of the extra Coulomb potential and diminish the effective strength and range of this Coulomb potential. The induced change in the density $\langle \delta\rho \rangle$ is called the Coulomb hole because electrons are pushed away from the added charge as shown in Fig. 2.3. The effective potential is the screened Coulomb potential W

$$W(12) = \int d3 V(13)\epsilon^{-1}(32). \tag{2.134}$$

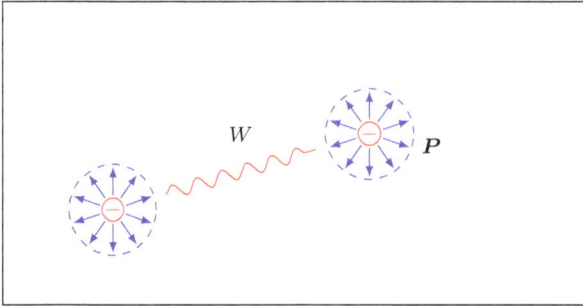

Fig. 2.3   Two negative test charges surrounded by a cloud of polarization $\boldsymbol{P}$. A charge with its polarization cloud constitutes a quasiparticle. The two quasiparticles interact through the screened Coulomb interaction W.

It is much more practical to reformulate the excited states of an electronic system into a collection of particles that interact through the screened Coulomb potential W instead of the bare Coulomb potential V because the screened potential tends to be short-ranged, especially in metals where the electrons can easily move and screen potential fluctuations. Electrons only interact via the bare Coulomb potential V, so the particles that the system describes are quasiparticles, or particles involving the collective motion of many electrons. The GW method, described later in section 2.1.7, is a Green's function method, making use of W, for determining the properties of the quasiprticle states.

2.1.6.3   *Local fields*

The susceptibility should describe the response of the system to fields that fluctuate on a length scale smaller than the unit cell. Waves with wavelengths smaller than the unit cell are best described by RLVs, so the Fourier transform of the susceptibility is

$$\chi(\boldsymbol{q}+\boldsymbol{G},\boldsymbol{q}+\boldsymbol{G}') = \iint \mathrm{d}\boldsymbol{r}\mathrm{d}\boldsymbol{r}' \mathrm{e}^{\mathrm{i}(\boldsymbol{q}+\boldsymbol{G})\cdot\boldsymbol{r}} \mathrm{e}^{\mathrm{i}(\boldsymbol{q}+\boldsymbol{G}')\cdot\boldsymbol{r}'} \chi(\boldsymbol{r},\boldsymbol{r}') \qquad (2.135)$$

where $\boldsymbol{q}$ is a wave vector within the first BZ. In this form of the susceptibility, the off-diagonal matrix elements $\chi(\boldsymbol{q}+\boldsymbol{G},\boldsymbol{q}+\boldsymbol{G}')$, with $\boldsymbol{G} \neq \boldsymbol{G}'$, represent nonuniform density fluctuations within the unit cell. The local fields are short-ranged electric fields caused by the redistribution of charge on the scale of the unit cell and are more prominent in insulators and semiconductors than in metals. Simple metals, that is, metals well-described by the nearly-free electron model (Li, Na, Al, ...), do not easily polarize within the unit cell because charge can easily flow to the edges of the bulk material. Northrup *et al.* (1989) calculated the local-field contributions to Li, Na and Al susceptibilities and found the off-diagonal matrix elements had little effect on the band structure in these materials. Insulators obtain the largest effect from the local fields in the bonding regions where the density is also localized. In insulators and semiconductors, an applied field polarizes the medium, due to small displacements of the electron charge density within the unit cell, and contributing to the off-diagonal matrix elements of the susceptibility. Hybertsen and Louie (1986) found that in Si, the local field contribution to the screening potential is around $-4\,\mathrm{eV}$ in the bonding region and from about -1.5 to $0.5\,\mathrm{eV}$ in the interstitial region.

2.1.7   *The GW method*

Hedin (1965) developed an *ab initio* method, based on many-body perturbation theory, that accurately describes excited states of atoms, molecules and solids using quasiparticle states. His main goal was to develop a perturbation expansion of the many-body interactions in terms of the screened

Coulomb interaction. His GW method closely relates to the idea of the simple angle-resolved photoemission spectroscopy (ARPES) for a sample containing N electrons. In the photoemission experiment an incoming photon ejects an electron from the valence band leaving the system with $N-1$ electrons. In the inverse photoemission experiment, an electron strikes the material, ejects a photon, increases the number of electrons $N+1$. The photoemission experiments probes the valence band and the excited states and is often used to determine the fundamental gap energy E_g.

Quasiparticles interact with each other through the screened Coulomb potential W, also shown in Fig. 2.3, instead of the bare Coulomb potential V. The quasiparticles are not quite eigenstates of the many-body Hamiltonian, but have complex eigenvalues with the imaginary part inversely proportional to the quasiparticle lifetime. Using the photoemission experiment as a guide, the binding energy of a quasiparticle state is the difference between an N-particle ground state Ψ_0^N and an excited state in a system of $N-1$ particles, Ψ_i^{N-1} such that $\epsilon_i = E_0^N - E_i^{N-1}$ when $\epsilon_i < \mu$, where μ is the chemical potential. Similarly, the inverse photoemission process adds an electron to the N-particle ground state, so $\epsilon_i = E_i^{N+1} - E_0^N$ when $\epsilon_i > \mu$. The quasiparticle states are

$$\Psi_i(\boldsymbol{r}, t) = \left\langle N, 0 \,\middle|\, \hat{\Psi}(\boldsymbol{r}, t) \,\middle|\, N+1, i \right\rangle \tag{2.136}$$

$$= \left\langle N+1, i \,\middle|\, \hat{\Psi}^\dagger(\boldsymbol{r}, t) \,\middle|\, N, 0 \right\rangle, \tag{2.137}$$

$$\epsilon_i = E_{N+1,i} - E_{N,0}, \text{ for } \epsilon_i \geq \mu, \text{ and} \tag{2.138}$$

$$\Psi_i(\boldsymbol{r}, t) = \left\langle N-1, i \,\middle|\, \hat{\Psi}(\boldsymbol{r}, t) \,\middle|\, N, 0 \right\rangle \tag{2.139}$$

$$= \left\langle N, 0 \,\middle|\, \hat{\Psi}^\dagger(\boldsymbol{r}, t) \,\middle|\, N-1, i \right\rangle, \tag{2.140}$$

$$\epsilon_i = E_{N,0} - E_{N-1,i}, \text{ for } \epsilon_i < \mu \tag{2.141}$$

where $\hat{\Psi}(\boldsymbol{r}, t)$ is the fermion field annihilation operator, which destroys a fermion located at $\boldsymbol{r}$ and time t.

The namesake for the GW method comes from the approximate expression for the quasiparticle self-energy operator Σ which involves the energy convolution of the Green function G with the screened Coulomb potential W

$$\Sigma(\boldsymbol{r}, \boldsymbol{r}', \omega) = \mathrm{i} \int d\omega' \, G(\boldsymbol{r}, \boldsymbol{r}', \omega + \omega') W(\boldsymbol{r}, \boldsymbol{r}', \omega'), \tag{2.142}$$

or in the short-hand notation

$$\Sigma = \mathrm{i}GW. \tag{2.143}$$

GW is the lowest order term in an infinite series of Feynman diagrams for the quasiparticle self-energy. The self-energy operator describes all of the non-local, energy dependent, quasiparticle interactions, including correlation and exchange, beyond the Hartree interaction. The quasiparticles obey the quasiparticle Hamiltonian

$$\hat{H}\Psi_i(1) = \left[-\frac{1}{2}\nabla_1^2 + v_{\text{ext}}(1) + v_{\text{H}}\right]\Psi_i(1) + \int d2\Sigma(12)\Psi_i(2) = \epsilon_i\Psi_i(1).$$

(2.144)

The self-energy is directly related to how the system responds to the addition or subtraction of a particle and, in this sense, it is related to the response function discussed previously. In order to solve for the quasiparticle wave functions, the goal is to find an appropriate approximation for Σ.

2.1.7.1 *Hedin's equations*

Hedin (1965) developed the GW method by linking 5 many-body equations together into a self-consistent scheme:

Dyson's equation	$G(12) = G_0(12) + G_0(13)\Sigma(34)G(42)$	(2.145)
Screening equation	$W(12) = v(12) + v(13)P(34)W(42)$	(2.146)
Self-energy	$\Sigma(12) = iW(1^+3)G(14)\Gamma(42;3)$	(2.147)
Polarizability	$P(1,2) = -iG(23)G(42^+)\Gamma(34;1)$	(2.148)
Vertex Correction	$\Gamma(12;3) = \delta(12)\delta(13)$	

$$+ \Gamma(14;7)G(64)G(78)\frac{\delta\Sigma(23)}{\delta G(68)}. \quad (2.149)$$

The Hedin equations are often summarized in the diagram in Fig. 2.4(a). The vertex correction, Eq. 2.149, arises from the two-particle Green's function associated with the Coulomb energy and it provides many-body corrections to the exchange and correlation energy. The vertex correction Γ is difficult to deal with numerically, and they tend to average out (DuBois, 1959a,b), so the GW approxmation (GWA) is employed such that $\delta\Sigma/\delta G = 0$. Figure 2.4(b) shows the path of the GWA where the complicated vertex terms are ignored. Figure 2.4(c) shows the "one-shot" G_0W_0 approximation which is the simplest form of the GWA. The GWA for the self energy is often represented by the Feynman diagram in Fig. 2.5.

By design, the GW method approximates the self-energy of the Hartree approximation, but it is also possible to start with any approximation (for

(a) GW

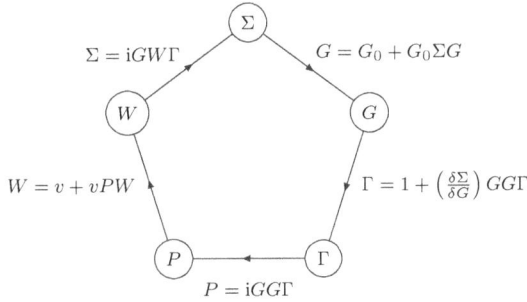

$$\Sigma = iGW\Gamma$$

$$G = G_0 + G_0\Sigma G$$

$$W = v + vPW$$

$$\Gamma = 1 + \left(\frac{\delta\Sigma}{\delta G}\right)GG\Gamma$$

$$P = iGG\Gamma$$

(b) GWA

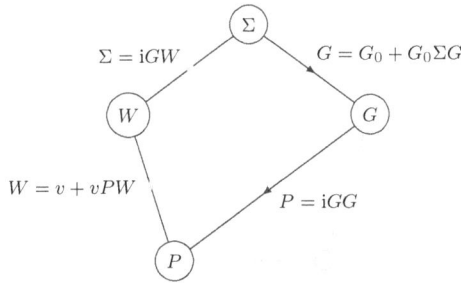

$$\Sigma = iGW$$

$$G = G_0 + G_0\Sigma G$$

$$W = v + vPW$$

$$P = iGG$$

(c) G_0W_0

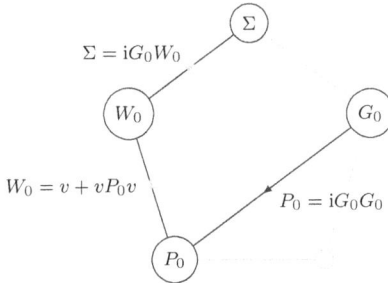

$$\Sigma = iG_0W_0$$

$$W_0 = v + vP_0v$$

$$P_0 = iG_0G_0$$

Fig. 2.4   Three types of GW methods. (a) The full GW procedure includes the compli-cated vertex terms Γ in the determination of the polarizability P and the self-energy Σ. The loop repeats until the convergence criteria is met. (b) The GW Approximation, or GWA, ignores the vertex correction and replaces it with a delta function. Like the GW method, the loop repeats. (c) The G_0W_0 approximation only completes one iteration of the loop starting with some appropriate G_0.

example, LDA, GGA, exact exchange (EXX), or local-density approxima-

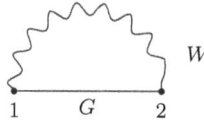

Fig. 2.5   Feynman diagram for the GW approximation. W is the screened Coulomb interaction between propagating quasiparticles and G is the free quasiparticle propagator.

tion with Hubbard U (LDA+U)) by solving the modified Hamiltonian

$$\epsilon_i \Psi_i(1) = h(1)\Psi_i(1) + v_H(1)\Psi_i(1) + v_{xc}(1)\Psi_i(1)$$

$$+ \int d2 \underbrace{[\Sigma(2) - \delta(12)v_{xc}(2)]}_{\Delta\Sigma(12)} \Psi_i(2). \qquad (2.150)$$

Generally, the better the starting v_{xc}, the better the G_0W_0 method works, because $\Delta\Sigma$ is small.

2.1.7.2   *The COHSEX approximation*

The convolution of G and W in the calculation of Σ can be time-consuming, so we will describe two approximations to the integral. The first approximation is the Coulomb-hole plus screened exchange (COHSEX) approximation.

The self-energy in the GWA involves the frequency convolution of G and W. Numerically, the convolution is difficult because response function would need to be evaluated at many points in ω-space to correctly describe the singularity near $\hbar\omega = \pm(E_m - E_n)$. Originally, Hedin (1965) avoided some of the complications of the convolution by analyzing the spectral properties of the self-energy. Generally, the main contribution to the self-energy comes from the Coulomb integral and the exchange integral involving only states with similar energies, and the self-energy can be split into two terms

$$\Sigma(12) = \frac{1}{2}\delta(12)W_p(12; \omega = 0) - W(12; \omega = 0)\sum_i \phi_i(\mathbf{r}_1)\phi_i^*(\mathbf{r}_2) \quad (2.151)$$

where W_p is $W-v$, the screened Coulomb potential minus the bare Coulomb potential and $\phi_i(\mathbf{r})$ are the independent particle states. W is evaluated at $\omega = 0$ because the main contribution to the screened potential is from transitions of similar energy. The first term in Eq. 2.151 corresponds to the Coulomb hole since it involves only the induced field from LR theory. The second term is the screened exchange (SEX) since it is subtracting the over-counting of energy in the Coulomb interaction, just as the exchange integral is subtracted form the direct Coulomb interaction in the HF energy. Together, these two terms in Eq. 2.151 constitute the COHSEX approximation.

2.1.7.3 *Plasmon-pole Approximations*

An alternative approximation to the calculation of Σ involves analytic forms, or models, of ϵ^{-1}. The idea is to analytically integrate $\Sigma = GW$ according to the model and fit to a few calculated frequencies. Plasmon-pole models (PPMs) are analytic functions that are fit to the plasmon behavior near the plasma frequency $\omega = \omega_p$ and in the static limit $\omega \to 0$.

Hybertsen and Louie (1985) observed that the imaginary part of the inverse susceptibility is generally a peaked function at some frequency $\pm\tilde{\omega}(q + G, q + G')$. They developed the generalized plasmon-pole model (GPPM) based on this observation using

$$\Im\epsilon^{-1}(q; G, G'; \omega) = A(q + G, q + G') \left[\delta(\omega - \tilde{\omega}) - \delta(\omega + \tilde{\omega})\right], \quad (2.152)$$

as the imaginary inverse susceptibility where A is the spectral weight (related to the rate of absorption/dissipation of plasmon modes). The real part of the function, which is related to the fluctuation of the plasmon modes, is determined from the Kramers–Kronig (KK) relation and the values of A and $\tilde{\omega}$ are completely determined by the Johnson sum rules for the moments of ϵ^{-1} (Johnson, 1974)

$$\int_0^\infty d\omega\, \omega \Im\epsilon^{-1}(q + G, q + G'; \omega) = -\frac{\pi}{2}\omega_p^2 \frac{\rho(G - G')}{\rho(0)} \frac{(q + G) \cdot (q + G')}{|q + G|^2}$$

$$(2.153)$$

where ρ is the charge density for the Fourier components.

Some of the first *ab initio GW* calculations, by Hybertsen and Louie (1986), showed that the *GW* method improves the gap of semiconductors and insulators over the LDA. For example, the *GW* gap using the GPPM on Si is 1.21 eV, compared to 0.52 eV with the LDA. The experimental gap is 1.17 eV. The GPPM for the screening only requires two susceptibility calculations for each $(q + G, q + G')$: one in the static limit and another near ω_p. The two calculations are enough, in combination with Eq. 2.153, to determine the model inverse dielectric function.

2.1.8 *Methods of calculating T_C*

The basic definition of the Curie temperature T_C is the temperature, above which, spins of a ferromagnetic system are no longer correlated. A simple microscopic picture of determining T_C is to consider the configurations and energies of spin fluctuations in a single-domain ferromagnet. At $T \approx 0\,\mathrm{K}$, all of the spins are aligned so the magnetization of the sample is the largest. As the thermal energy increases, thermal fluctuations reorient the spins decreasing the overall magnetization. At T_C, the random thermal fluctuations of the spins overcome the exchange interactions that cause them to align. At temperatures above T_C, we expect that the behavior of the magnetization to follow the Curie-Wiess law

$$\chi = \left.\frac{\partial M}{\partial h}\right|_{h=0} = \frac{C}{T - T_C} \tag{2.154}$$

where χ is the susceptibility and C is the Curie constant. Below T_C, M is finite at $h = 0$ (spontaneous magnetization). At temperatures well above T_C, the interactions between moments should be very small compared to the thermal energy, so we expect the system to behave as a paramagnet, that is, the susceptibility follows the Curie law for paramagnetism

$$\chi = \frac{C}{T}. \tag{2.155}$$

In general, T_C is found by developing an expression for χ and comparing it to Eq. 2.154. There are a number of methods, based on DFT, that can be implemented to calculate χ and T_C by first calculating the exchange parameter J_{ij} used in the Heisenberg Hamiltonian. After discussing how to obtain J_{ij} from DFT calculations, we will discuss two methods, using two different mathematical frameworks, to obtain T_C. The methods are the mean-field approximation (MFA) and the random phase approximation (RPA).

2.1.8.1   *Determining J_{ij}*

The exchange parameter J_{ij} can be determined from the total energy calculations from DFT by considering the total energy of the spin-spiral structure in Fig. 2.6. Using the Heisenberg Hamiltonian, the total energy associated with the spin spiral is

$$E_{\mathrm{DFT}}(\boldsymbol{q}, \theta) = E_0 - \frac{1}{2}\sum_{ij} J_{ij} \boldsymbol{S}_i \cdot \boldsymbol{S}_j \tag{2.156}$$

where E_0 is the part of the total energy that is not dependent on the spin state. From the schematic of the spin spiral structure, the two vectors $\boldsymbol{S}_i$ and $\boldsymbol{S}_j$ have cartesian components

$$\boldsymbol{S}_i = S\left(\sin\theta\cos\phi, \sin\theta\sin\phi, \cos\theta\right) \tag{2.157}$$

$$\boldsymbol{S}_j = S\left(\sin\theta\cos(\phi + \boldsymbol{q}\cdot\boldsymbol{R}_{ij}), \sin\theta\sin(\phi + \boldsymbol{q}\cdot\boldsymbol{R}_{ij}), \cos\theta\right) \tag{2.158}$$

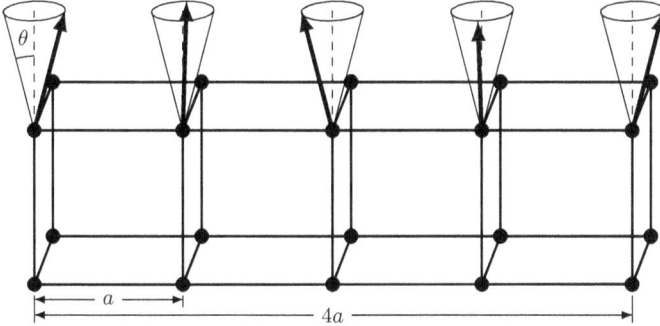

Fig. 2.6   Spin spiral structure with $\boldsymbol{q} = (\pi/2a)\hat{\boldsymbol{x}}$. The thick arrows indicate the spin direction at each site.

where $\boldsymbol{R}_{ij} = \boldsymbol{R}_j - \boldsymbol{R}_i$. The phase factor ϕ is arbitrary, so it is set to zero, and the energy is

$$E(\boldsymbol{q}) = E_0 - \frac{1}{2}\sum_{ij} J_{ij} S^2 \left(\sin^2\theta \cos(\boldsymbol{q}\cdot\boldsymbol{R}_{ij}) + \cos^2\theta\right) \tag{2.159}$$

The energy difference between the spin spiral structure and the ferromagnetic one ($\boldsymbol{q} = 0$) is

$$E_{\mathrm{DFT}}(\boldsymbol{q},\theta) - E_{\mathrm{DFT}}(\boldsymbol{q}=0,\theta) = \frac{1}{2}S^2\sin^2\theta\sum_{ij} J(\boldsymbol{R}_{ij})[\cos(\boldsymbol{q}\cdot\boldsymbol{R}_{ij}) - 1] \tag{2.160}$$

Using the Fourier transform (FT) of J_{ij}

$$J(\boldsymbol{R}_i - \boldsymbol{R}_j) = \frac{1}{N}\sum_{\boldsymbol{q}} J(\boldsymbol{q})e^{i\boldsymbol{q}\cdot(\boldsymbol{R}_i - \boldsymbol{R}_i)}, \tag{2.161}$$

the energy difference is

$$E_{\mathrm{DFT}}(\boldsymbol{q},\theta) - E_{\mathrm{DFT}}(\boldsymbol{q}=0,\theta) = \frac{1}{2}S^2\sin^2\theta[J(\boldsymbol{q}) - J(\boldsymbol{q}=0)]. \tag{2.162}$$

Sandratskii (1998) introduced a very important idea to carry out spin spiral calculations without using supercells. The generalized translation vector for a spin spiral structure is

$$P_{\boldsymbol{k}}\Psi_k = U(\alpha_S)e^{-i\boldsymbol{k}\cdot\boldsymbol{R}_n}\Psi_k \tag{2.163}$$

where $U(\alpha_S)$ only acts on the spin part of the wave function

$$U(\alpha_S) = \begin{pmatrix} e^{-i\boldsymbol{q}\cdot\boldsymbol{R}_i/2} & 0 \\ 0 & e^{i\boldsymbol{q}\cdot\boldsymbol{R}_i/2} \end{pmatrix}. \tag{2.164}$$

For each $\boldsymbol{k}$-point, the eigenvalue of the nonmagnetic system $\epsilon(\boldsymbol{k})$ is split into to $\epsilon(\boldsymbol{k}\pm\boldsymbol{q}/2)$. $E_{\mathrm{DFT}}(\boldsymbol{q})$ can be obtained by summing the eigenvalues and then subtracting $E_{\mathrm{DFT}}(\boldsymbol{q}=0)$. From E_{DFT} one can get $J(\boldsymbol{q})$ from Eq. 2.162.

2.1.8.2 *Thermodynamic background*

In the previous methodological sections, such as DFT and the GW method, we have neglected to discuss temperature. Typically, these methods are utilized to determine ground state and low energy excitations through linear response. To understand T_C, we need to introduce thermal configurations of states that cause the magnetization M of a ferromagnetic spintronic sample to decrease to zero.

The following general Hamiltonian is parameterized by the strength of some interaction Hamiltonian

$$H_\eta = H_0 + \eta H' \tag{2.165}$$

where H_0 is the zeroth order Hamiltonian which can easily be solved, H' is some Hamiltonian characterizing some interaction indicated by the strength parameter η, $0 \leq \eta \leq 1$. Later, we will use the Zeeman term for H_0 and the Heisenberg Hamiltonian for H'. Temperature is introduced in the form of the partition function Z_η and free energy F_η

$$Z_\eta = \text{Tr} \exp(-\beta H_\eta) \tag{2.166}$$

$$F_\eta = -\frac{1}{\beta} \ln Z_\eta \tag{2.167}$$

where $\beta = 1/k_B T$, k_B is the Boltzmann constant, and Tr is the trace over configuration states of H_0, or any complete set of states describing the system.

Thermodynamic equilibrium of the full system is determined by minimizing the free energy F_η with respect to the parameter η. The derivatives of F_η are

$$\frac{\partial F_\eta}{\partial \eta} = -\frac{1}{\beta} \frac{1}{Z_\eta} \frac{\partial Z_\eta}{\partial \eta} \tag{2.168}$$

$$= \langle H' \rangle_\eta \tag{2.169}$$

$$\frac{\partial^2 F_\eta}{\partial \eta^2} = -\beta \left[\langle H'^2 \rangle_\eta - \langle H' \rangle_\eta^2 \right] \leq 0, \tag{2.170}$$

where the thermodynamic average operator is defined as

$$\langle X \rangle_\eta = \frac{1}{Z_\eta} \text{Tr} \left[X \exp \left(-\beta H_\eta \right) \right] \tag{2.171}$$

and X can be any operator of the system. From the Taylor expansion of F_η, and using the fact that the second derivative of F_η with respect to η is strictly negative,

$$F_\eta \leq F_0 + \eta \left[\frac{\partial F_\eta}{\partial \eta} \bigg|_{\eta=0} \right], \tag{2.172}$$

so the minimum value of $F_{\eta=1}$ is

$$F_1 = F_0 + \langle H' \rangle_{\eta=0} \tag{2.173}$$

$$F_1 = F_0 + \langle H_1 - H_0 \rangle_0 \tag{2.174}$$

2.1.8.3 Zeroth order solution (Paramagnetism)

The zeroth order Hamiltonian is the interaction of the effective field $h\hat{z}$ acting on each magnetic moment $\boldsymbol{\mu}_i$

$$H_0 = -h \sum_i \hat{z} \cdot \boldsymbol{\mu}_i \tag{2.175}$$

where $\hat{z}$ is the unit vector in the z-direction.

The zeroth order partition function from the Hamiltonian in Eq. 2.175 is easily solved in both the classical and quantum limits because each dipole only interacts with the external field and not each other. In the quantum limit, $\boldsymbol{\mu}_i$ is replaced by $g\mu_B \mathbf{S}_i$, where g is the g-factor, the z-component S_i^z can only take values $(-S, -S+1, \ldots, S)$. Additionally, since there is no coupling between the spins, each site has its own partition function. Thus, the partition function at $\eta = 0$ is

$$Z_0 = \left[\sum_{m=-S}^{S} \exp\left(-g\mu_B h \beta m\right) \right]^N \tag{2.176}$$

$$= \left[\frac{\sinh\left\{ \left(1 + \frac{1}{2S}\right) y \right\}}{\sinh\left\{ \left(\frac{1}{2S}\right) y \right\}} \right]^N \tag{2.177}$$

where $y = \beta g \mu_B h S$ is the reduced spin factor, and N is the total number of magnetic moments. The energy $\langle H_0 \rangle_0$ is

$$\langle H_0 \rangle_0 = -\frac{\partial}{\partial \beta} \ln Z_0 \tag{2.178}$$

$$= -hM \tag{2.179}$$

$$M = -\frac{1}{\beta} \frac{\partial}{\partial h} \ln Z_0 \tag{2.180}$$

$$= N g \mu_B S B_S(y) \tag{2.181}$$

where B_S is the Brillouin function

$$B_S(y) = \left(1 + \frac{1}{2S}\right) \coth\left\{ \left(1 + \frac{1}{2S}\right) y \right\} - \frac{1}{2S} \coth\left\{ \frac{y}{2S} \right\}. \tag{2.182}$$

At this point, we may investigate the high temperature regime of the susceptibility to recover Curie's law of paramagnetism (Eq. 2.155). As $y \to 0$, $B_S(y) \to \frac{y}{3}(1 + 1/S)$, so

$$\chi_C = \left.\frac{\partial M}{\partial h}\right|_{h=0} = \frac{C}{T} \tag{2.183}$$

and the Curie constant is

$$C = \frac{N(g\mu_B)^2 S(S+1)}{3k_B}. \tag{2.184}$$

2.1.8.4 *Mean-field approximation to T_C*

For a simple derivation using the MFA, we consider that the physical origin of T_C is due to the average effective magnetic field, the Weiss molecular field, due to the neighboring magnetic moments, acting on a magnetic moment. For the perturbing Hamiltonian, we will use the Heisenberg Hamiltonian acting over nearest neighbors only

$$H' = -\frac{1}{2} \sum_{ij} J_{ij} \boldsymbol{S}_i \cdot \boldsymbol{S}_j \tag{2.185}$$

where J_{ij} is the positive exchange function that depends on the vector $\boldsymbol{R}_j - \boldsymbol{R}_i$ from site i to site j.

In the MFA, a spin at site i experiences the average field due to the remaining $N - 1$ sites. The dot product $\boldsymbol{S}_i \cdot \boldsymbol{S}_j$ should be replaced by an interaction between $\boldsymbol{S}_i$ and the average $\langle \boldsymbol{S}_j \rangle$

$$H' = -\frac{1}{2} \sum_i \boldsymbol{S}_i \cdot \left(\sum_j J_{ij} \boldsymbol{S}_j \right) \tag{2.186}$$

$$= -\sum_i \boldsymbol{\mu}_i \cdot \boldsymbol{h}_i^{\text{eff}} \tag{2.187}$$

where $\mu_i = g\mu_B S_i$ is the magnetic moment of spin i and we have defined the effective field at site i

$$\boldsymbol{h}_i^{\text{eff}} = \frac{1}{2g\mu_B} \left(\sum_j J_{ij} \boldsymbol{S}_j \right). \tag{2.188}$$

We make the assumption that the effective field is identical at every site and points in the $\hat{\boldsymbol{z}}$-direction, so the MFA hamiltonian is identical to the zeroth order Hamiltonian, but with an effective field instead of an externally applied field h:

$$H' = -h_{\text{eff}} \sum_i \hat{\boldsymbol{z}} \cdot \boldsymbol{\mu}_i \tag{2.189}$$

where

$$h_{\text{eff}} = \frac{M}{2Ng^2\mu_B^2} \left(\sum_j J_{ij} \right) \tag{2.190}$$

$$M = Ng\mu_B \langle S \rangle . \tag{2.191}$$

Next, we replace h by $h + h_{\text{eff}}$ in the paramagnetic solution for M (Eq. 2.181)

$$M = Ng\mu_B S B_S \left[\beta g\mu_B (h + \lambda M)S \right], \text{ where} \tag{2.192}$$

$$\lambda = \frac{1}{2N(g\mu_B)^2} \sum_j J_{ij}, \tag{2.193}$$

and solve the equation

$$\left. \frac{dM}{dh} \right|_{h=0} = \frac{C}{T - T_C} \tag{2.194}$$

for the Curie temperature

$$T_C = \frac{C}{2N(g\mu_B)^2} \sum_j J_{ij}. \tag{2.195}$$

2.1.8.5 *Random-phase approximation*

The RPA was originally examined by Bohm and Pines (1951) to describe the collective oscillations of electrons called plasmons. The basic approximation is to assume that the response of the system at different wavelengths tend to average to zero. That is, if the system interacts with an external field with wavelength λ, then the fluctuations of the system will also have wavelength λ in the same direction.

As a specific example of the RPA, we consider the derivation of the plasmon frequency within the RPA. While the response function formalism seen in section 2.1.6 is well-suited for demonstrating the RPA, we instead follow the second quantization method. Using second quantization, the Hamiltonian for a collection of electrons experiencing a mutual Coulomb repulsion is

$$H = \sum_k \epsilon_k c_k^\dagger c_k + \sum_k \sum_{k'} \sum_{q \neq 0} \left(\frac{4\pi e^2}{q^2} \right) c_{k+q}^\dagger c_{k'-q}^\dagger c_{k'} c_k \tag{2.196}$$

where ϵ_k is the single particle energy of an electron with momentum $\hbar k$, $2\pi e^2/q^2$ is the Fourier transform of the Coulomb potential, and c_k ($c_k^\dagger$) is the fermion annihilation (creation) operator for particles with momentum $\hbar k$. The density operator and its equations of motion are

$$\rho_q^\dagger = \rho_{-q} \tag{2.197}$$

$$= \sum_k c_{k+q}^\dagger c_k \tag{2.198}$$

$$i\hbar \frac{\partial \rho_q^\dagger}{\partial t} = \left[H, \rho_q^\dagger\right] \tag{2.199}$$

$$= \sum_k (\epsilon_{k+q} - \epsilon_k) \, c_{k+q}^\dagger c_k \tag{2.200}$$

$$i\hbar \frac{\partial}{\partial t} \left(i\hbar \frac{\partial \rho_q^\dagger}{\partial t} \right) = \left[H, \left[H, \rho_q^\dagger\right]\right] \tag{2.201}$$

$$= \sum_k (\epsilon_{k+q} - \epsilon_k)^2 \, c_{k+q}^\dagger c_k + \left[V, \left[H, \rho_q^\dagger\right]\right] \tag{2.202}$$

The calculations of the commutators for the second time derivatives are tedious, but the result, when using ϵ_k as the kinetic energy and assuming the form $\rho_q^\dagger \sim \exp(-i\omega t)$, is

$$\hbar^2 \omega^2 \rho_q^\dagger = \sum_k \left[\frac{\hbar^2}{2m} (2q \cdot k + q^2) \right]^2 c_{k+q}^\dagger c_k \tag{2.203}$$

$$+ \sum_{q'} \left[V_{q'} \frac{\hbar^2}{2m} q' \cdot q \right] \left(\rho_{q'-q} \rho_{q'}^\dagger + \rho_{-q'}^\dagger \rho_{-q'+q} \right) \tag{2.204}$$

where $V_q = 2\pi e^2/q^2$. Consider the order of magnitude of these terms: the first term is of order q^4 and the second term is of order 1. For long wavelengths, the first term may be ignored so we will concentrate on the second term. For the second term, we employ the RPA and expect that the $q' \neq q$ terms tend to average to zero. The only remaining term from the sum is $q' = q$. The resulting equation of motion (EOM) is

$$\hbar^2 \omega^2 \rho_q^\dagger = \frac{2\pi e^2}{q^2} \frac{\hbar^2 q^2}{2m} (2\rho_0) \, \rho_q^\dagger. \tag{2.205}$$

Within the RPA, the plasmon frequency ω_P is just a constant that depends on the average density ρ_0

$$\omega_P = \sqrt{\frac{2\pi e^2 \rho_0}{m}}. \tag{2.206}$$

2.1.8.6 *Random phase approximation to T_C*

The RPA has been applied to solve many physical problems includeing the determination of T_C. The RPA method we choose to investigate is based on the two-time Green's function described by Bogolyubov and Tyablikov (1959). The Hamiltonian we use is the Zeeman and Heisenberg Hamiltonian

$$H = H_0 + H' \tag{2.207}$$

$$= -\mu_B h_0 \sum_i S_i^z - \frac{1}{2} \sum_{\langle ij \rangle} J_{ij} \boldsymbol{S}_i \cdot \boldsymbol{S}_j \tag{2.208}$$

where $\boldsymbol{h}_0$ is an external field, μ_B is the Bohr magneton and the factor of $1/2$ is from over-counting.

In the MFA, the second term in Eq. 2.208 is reduced to an effective field at any site i. The RPA treats the second term without the site averaging. In terms of Fourier components, there will be operators involving different values of $\boldsymbol{q}$ and $\boldsymbol{q}'$ but only $\boldsymbol{q}$ will not average to zero. The method described by Tyablikov (1959) truncates the higher order Green's functions that characterize the multiple scattering of spin waves, thus removing the dependence on $\boldsymbol{q}'$. As a consequence, this method is often called RPA in the literature. The derivation of the expression for T_C in the RPA is tedious so many authors simply quote the result. There are a few authors that provide general derivations. They are Callen (1963), Tahir-Kheli (1962) and Rusz *et al.* (2005). The basic descriptions in these papers are similar; however, there is variation in the notation. In the following, we demonstrate a simple derivation utilizing the two-time Green's functions to solve T_C for a spin-$1/2$ ($S = 1/2$) system. The Hamiltonian and the Green's function involve the spin operators with the following properties:

$$S_i^\pm = S_i^x \pm iS_i^y \tag{2.209}$$

$$[S_i^+, S_j^-] = 2S_j^z \delta_{ij} \tag{2.210}$$

$$[S_i^\pm, S_j^z] = \mp S_j^\pm \delta_{ij} \tag{2.211}$$

$$\boldsymbol{S}_i \cdot \boldsymbol{S}_j = S_i^x S_j^x + S_i^y S_j^y + S_i^z S_j^z$$

$$= S_i^z S_j^z + \frac{1}{2} \left[S_i^+ S_j^- + S_i^- S_j^+ \right] \tag{2.212}$$

$$\boldsymbol{S}_i \cdot \boldsymbol{S}_i = (S_i^z)^2 + S_i^z + S_i^- S_i^+$$

$$= S(S + 1) \tag{2.213}$$

where we have let $\hbar = 1$ for simplicity. Additionally, for spin 1/2 systems,

$$(S_i^+)^2 = (S_i^-)^2 = 0, \text{ and} \tag{2.214}$$

$$(2S_i^z + 1)(2S_i^z - 1) = 0, \text{ so} \tag{2.215}$$

$$(S_i^z)^2 = 1/4. \tag{2.216}$$

The central quantity in this method is the correlation function given by Zubarev (1960)

$$\langle B(t')A(t) \rangle = \lim_{\eta \to 0^+} i \int dE \, \frac{\langle\!\langle A; B \rangle\!\rangle_{E+i\eta}^r - \langle\!\langle A; B \rangle\!\rangle_{E-i\eta}^r}{e^{\beta E} - 1} e^{-iE(t-t')} \tag{2.217}$$

where $\langle\!\langle A; B \rangle\!\rangle_E^r$ denotes the FT of the retarded two-time Green's function

$$G^r(t, t') = \langle\!\langle A(t); B(t') \rangle\!\rangle^r \tag{2.218}$$

$$= -i\theta(t - t') \langle [A(t), B(t')] \rangle \tag{2.219}$$

where $\langle A \rangle$ is defined previously in Eq. 2.171 and $\theta(t)$ is the step function that steps from zero to unity when the argument passes from negative to positive. The time dependence of operators in the Heisenberg picture is

$$A(t) = e^{-iHt} A(0) e^{iHt} \tag{2.220}$$

where $A(0) = A$ is the time-independent operator in the Schrödinger picture. The thermal average involves the trace over states, so it is straightforward to show that the retarded Green's function only depends on $\tau = t - t'$:

$$G^r(\tau) = \langle\!\langle A(\tau); B \rangle\!\rangle^r. \tag{2.221}$$

Equation 2.213 provides an expression relating the correlation function $\langle S_i^- S_i^+ \rangle$ to the total spin, so we can utilize Eq. 2.217 to determine $\langle S_i^- S_j^+ \rangle$ by solving the FT of the Green's function. The EOM for this Green's function is

$$i\frac{d}{dt} \langle\!\langle S_i^+(\tau); S_j^- \rangle\!\rangle^r = \frac{d\theta(\tau)}{dt} \langle [S_i^+(\tau), S_j^-] \rangle$$

$$+ \theta(\tau) \left\langle \left[\frac{dS_i^+(\tau)}{dt}, S_j^- \right] \right\rangle \tag{2.222}$$

$$= \delta(\tau) \langle [S_i^+(\tau), S_j^-] \rangle$$

$$+ i\langle\!\langle [S_i^+(\tau), H]; S_j^- \rangle\!\rangle^r, \tag{2.223}$$

and using the inverse FT and forward FT on τ,

$$G^r(\tau) = \int_{-\infty}^{\infty} dE \, G^r(E) e^{-iE\tau} \tag{2.224}$$

$$G^r(E) = \langle\!\langle S_i^+; S_j^- \rangle\!\rangle_E^r \tag{2.225}$$

$$= \frac{1}{2\pi} \int_{-\infty}^{\infty} d\tau \, G^r(\tau) e^{iE\tau}, \tag{2.226}$$

respectively, it becomes

$$EG^r(E) = \frac{1}{2\pi} \left\langle \left[S_i^+, S_j^- \right] \right\rangle + \langle\langle [S_i^+, H], S_j^- \rangle\rangle_E^r, \text{ where} \qquad (2.227)$$

$$\langle\langle [S_i^+, H], S_j^- \rangle\rangle_E^r = \frac{i}{2\pi} \int d\tau \, e^{iE\tau} \langle\langle [S_i^+(\tau), H] ; S_j^- \rangle\rangle^r \qquad (2.228)$$

is the FT of the second order Green's function. Next, inserting the Heisenberg Hamiltonian into Eq. 2.227 and following the commutator rules results in

$$(E - \mu_B h_0) \langle\langle S_i^+; S_j^- \rangle\rangle_E^r = \frac{\delta_{ij}}{\pi} \langle S^z \rangle$$
$$- \sum_k J_{ik} \langle\langle S_i^z S_k^+ - S_k^z S_i^+ ; S_j^- \rangle\rangle_E^r. \qquad (2.229)$$

The higher order Green's functions are recursive functions involving lower order Green's functions so the infinite series must be truncated depending on the physical situation. Following the *ad hoc* scheme by Tyablikov, the longitudinal (S^z) and transverse ($S^\pm$) motions are separated and S^z is replaced by its expectation value

$$\langle\langle S_i^+ S_k^z; S_j^- \rangle\rangle_E^r \overset{k \neq i}{\to} \langle S^z \rangle \langle\langle S_i^+; S_j^- \rangle\rangle_E^r. \qquad (2.230)$$

In a periodic solid, translational invariance allows FT the functions to real-space

$$G^r(E) = \frac{1}{N} \sum_q G^r(E, q) e^{iq \cdot (R_i - R_j)} \qquad (2.231)$$

$$\delta_{ij} = \frac{1}{N} \sum_q e^{iq \cdot (R_i - R_j)} \qquad (2.232)$$

where R_i is the i-th lattice vector and N is the number of q-points. The Fourier transform of the EOM is

$$(E - \mu_B h_0) G^r(E, q) = \frac{1}{\pi} \langle S^z \rangle - \langle S^z \rangle G^r(E, q) \sum_k J_{ik} \left[e^{iq \cdot (R_i - R_k)} - 1 \right].$$
$$(2.233)$$

Using the FT of the exchange

$$J(q) = \sum_k J_{ik} e^{-iq \cdot (R_i - R_k)}, \qquad (2.234)$$

the solution to the Green's function is

$$G^r(E, q) = \frac{\langle S^z \rangle}{\pi [E - \mu_B h_0 + \langle S^z \rangle (J(q) - J(0))]} \qquad (2.235)$$

$$\langle\langle S_i^+; S_j^- \rangle\rangle_E^r = \frac{1}{N} \sum_q \frac{\langle S^z \rangle}{\pi [E - E(q)]} e^{iq \cdot (R_i - R_j)} \qquad (2.236)$$

where $E(q) = \mu_B h_0 - \langle S^z \rangle [J(q) - J(0)]$. Using the decoupling scheme by Tyablikov (1959), the Green function is only dependent on q, which is precisely the condition for the RPA.

The time-dependent correlation function in Eq. 2.217 is now solvable:

$$S(S+1) - \langle S^z \rangle - \langle (S^z)^2 \rangle = \frac{2 \langle S^z \rangle}{N} \sum_q \frac{1}{e^{\beta E(q)} - 1}$$

$$= 2 \langle S^z \rangle \Phi(1/2) \tag{2.237}$$

In the above expression, we calculated the correlation at a single site $(i = j)$ at time $\tau = t - t' = 0$, let $h_0 \to 0$, and used the identity

$$\lim_{\eta \to 0^+} \left[\frac{1}{x + i\eta} - \frac{1}{x - i\eta} \right] = -2\pi i \delta(x) \tag{2.238}$$

for the Green function. Finally, we defined the function

$$\Phi(1/2) = \frac{1}{N} \sum_q \frac{1}{e^{\beta E(q)} - 1} \tag{2.239}$$

which is related to the number of spin excitations at temperature $T = 1/k_B \beta$. Rearranging Eq. 2.237, setting $S = 1/2$, and making use of the identity for $(S^z)^2$ in Eq. 2.216, the average spin moment is

$$\langle S^z \rangle = \frac{S(S+1) - \langle (S^z)^2 \rangle}{1 + 2\Phi(1/2)} \tag{2.240}$$

$$= \frac{1/2}{1 + 2\Phi(1/2)}, \tag{2.241}$$

which should tend toward zero as $T \to T_C$. In general, $\Phi(1/2)$ depends on $\langle S^z \rangle$ so the average spin moment below T_C must be solved self-consistently.

Since we are mainly interested in calculating T_C, we can take the limit as $\langle S^z \rangle$ approaches zero in Eq. 2.237 and solve for T_C. The exponential in the denominator of Eq. 2.237 expands and the factors of $\langle S^z \rangle$ cancel. The result is

$$\frac{1}{k_B T_C} = 4 \frac{1}{N} \sum_q \frac{1}{J(0) - J(q)} \tag{2.242}$$

$$= \frac{4}{J(0)} \frac{1}{N} \sum_q \frac{1}{1 - \gamma_q} \tag{2.243}$$

where $\gamma_q = J(q)/J(0)$.

2.2 Growth methods

In this section, the growth of Si-based spintronic materials is discussed. Desirable MR properties are often achieved by multilayered structures, so the ability to grow HM structures in thin-film form is of crucial importance. We review a number of experimental methods that can be used to dope TMEs in silicon. Nakayama *et al.* (2001) described that Mn, as well as many other TMEs, cannot be dissolved into Silicon at high enough concentrations for spintronic applications. Mn has a maximum solubility of $\sim 10^{16}/\text{cm}^2$ or $\sim 10^{-5}$ at.% (atomic percent) in crystalline silicon (Wiehl *et al.*, 1982). Doping concentrations of about 1 to 10 at.% are necessary for spintronic applications so non-equilibrium processes are necessary to obtain desirable properties. First, we discuss ion implantation, which has been used by Bolduc *et al.* (2005) to dope Mn into Si up to concentrations of 0.8 at.%. Next, we discuss a few methods where ions are evaporated into a vacuum and deposited on a substrate. These are called vacuum deposition methods and have been used by various groups to obtain concentrations as high as 15 at.%. Finally, we discuss molecular beam epitaxy (MBE), in which a beam of ions is deposited on a substrate under ultra high vacuum conditions.

During or after a sample is grown, it is necessary to characterize its quality and to measure its physical properties. In section 2.3, we will discuss the methods of characterizing thin films and determining various physical properties relevant to spintronic applications.

2.2.1 *The ion implantation method*

The ion implantation method dopes ions into a crystal by accelerating single species of ions from an electrode and bombarding them into the crystal surface. The typical setup has a source of ions, a section to select and accelerate the ions, and a wafer consisting of the crystal to be implanted. The schematic setup is shown in Fig. 2.7. The ion source chamber is located to the left. Ions are accelerated by the electric field provided by the accelerating electrode. The accelerated ion beam passes through a bending magnet to select ions of a specific mass. Any unwanted ions are deflected towards the wall of the mass selection tube. The remaining ions are then accelerated again by the "acceleration tube". Next, the ions pass through a focusing lens before hitting the target at bottom. A mask is often used to further control the positioning and penetration depth of the ions. Dosage can range from 10^{11} to 10^{16} ions per cm^2. The dosage, or fluence, is the total number of ions that pass through the the unit area. Channeling can occur if the beam is aligned with a major crystal direction, such as [100], and the beam can pass through many layers of the material before depositing the ion. To remedy the effect of channeling, the ion beam should be rotated away from the major crystal direction.

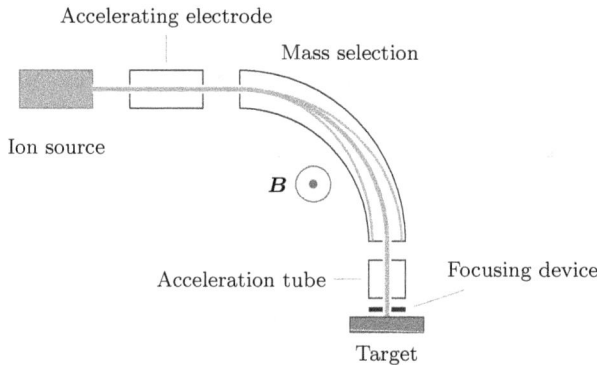

Fig. 2.7   Schematic setup of ion beam implantation.

2.2.1.1   *Ion beam energy and range*

The energy of the ion beam controls the dosage and penetration depth into the crystal. The energies range between 10 to 1000 keV. The penetration depth is called the "range". By adjusting the energy of the beam, the range can be controlled within 10 to 1000 nm.

2.2.1.2   *Energy loss*

The term "stopping" is used to characterize the energy loss of the ions within the crystal. At lower ion energy, or with lighter ions, energy loss is dominated by collisions with electrons while at higher ion energy, or heavier atoms, collisions with nuclei dominates. The type and rate of energy loss determines the range of the ion implantation.

2.2.1.3   *Crystal damage*

Implanting ions typically results in damage to the crystal structure, especially if the ions are heavy and energy loss is dominated by collisions with nuclei. Collisions with nuclei cause a cascade effect that generates vacancies and occupation of interstitial (I) sites. Understanding how ion implantation causes undesirable damage to the substrate is extremely important to minimize the effect. If the damage is sparse after ion implantation, annealing can restore the crystalline environment and the dopants can diffuse to their final positions. At high enough ion concentrations, however, the damage can be dense and crystals such as Si can become amorphous. Kucheyev *et al.* (2001) suggested that implanting ions at elevated substrate temperature can suppress the damage. In Si, heavy ion implantation dose bombardment can reach as high as $10^{15}/cm^2$ at RT before the Si substrate becomes amorphous.

2.2.2 *Vacuum deposition methods*

Vacuum deposition describes a class of methods that use the evaporation of a sources material to grow thin films on a substrate. It consists of a sources of atoms or molecules that are heated to form a gas in a low pressure vacuum chamber. The vacuum region serves three purposes:

- it increases the mean free path of the source particles,
- it controls the composition of the source particles, and
- it reduces the contaminants.

The gas solidifies on a target forming the thin-film or structure. It is possible to have layers ranging from a thickness of one atom up to millimeters. A schematic of a simple vacuum deposition device, using a heated ion source, is shown in Fig. 2.8. The square outline is the vacuum chamber. The target is a substrate for the thin film to be grown on and is located at the top of the chamber. The evaporator at the bottom emits the ionic vapor. Two vacuum deposition methods are detailed next.

2.2.2.1 *Pulsed laser deposition*

Pulsed laser deposition uses a laser to heat the source for ejecting ions. The schematic setup is shown in Fig. 2.9. A laser is focused through a focusing lens onto a source of ions at the bottom of the vacuum chamber. The laser evaporates the ion source and forms a plume of particles that are deposited on the target. The vacuum chamber increases the mean-free path of the particle plume. In principle, the ejected species can take the form of ions, atoms or even molecules before reaching the substrate.

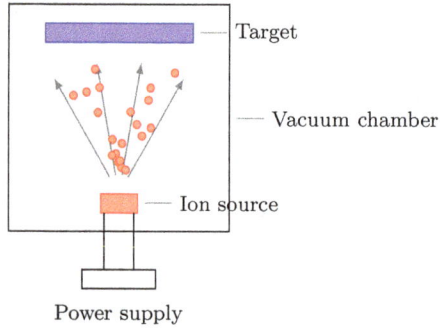

Fig. 2.8　A schematic setup for vacuum deposition.

Demidov *et al.* (2006) used this method to obtain Mn concentrations in Si from 10 to 15 at. %.

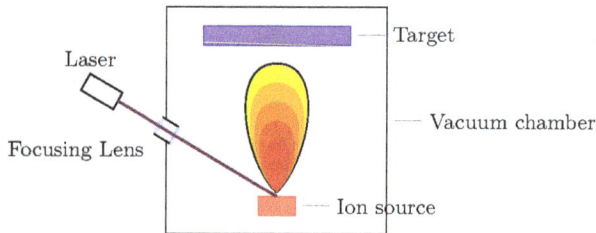

Fig. 2.9　Schematic setup of laser deposition.

2.2.2.2　*The chemical vapor deposition method*

The chemical vapor deposition (CVD) method uses a chemical precursor in gas form to deposit particles on or etch particles from a surface. A schematic setup for CVD is shown in Fig. 2.10. The substrate target is heated to assist the chemical reaction with the surface. The chemical precursor comes in from the intake, reacts with the target and then exits the chamber. Two important processes occur during the reaction with the target substrate: thermodynamic and kinetic. The thermodynamic process is characterized by the Gibbs free energy change

$$\Delta G = \Delta H - T\Delta S \qquad (2.244)$$

can either deposit ($\Delta G > 0$) or remove ($\Delta G < 0$) particles from the surface. H is the enthalpy associated with the chemical reaction, T is the substrate temperature, and S is the entropy. The kinematic process is much more complicated because it involves the transport of particles along the surface in the direction of the gas flow and the presence of nucleation sites or steps on the surface. The kinematic processes are detailed in Pattanaik and Sarin (2000).

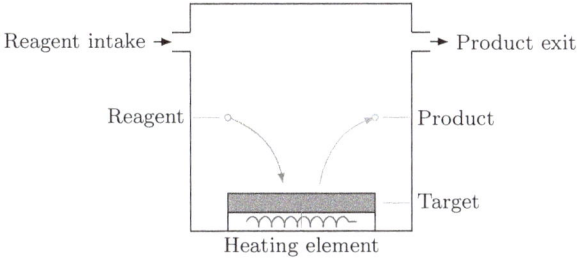

Fig. 2.10   Schematic of the chemical vapor deposition method.

2.2.3   *The molecular beam epitaxy method*

The MBE method is used to grow high-quality thin films. The schematic setup is given in Fig. 2.11. The setup includes a vacuum chamber where growth occurs on the sample, a rotating mounting device for the sample, four sources of atoms, one of which is shuttered, and a cryopanel or heating element to carefully control the temperature of the sample.

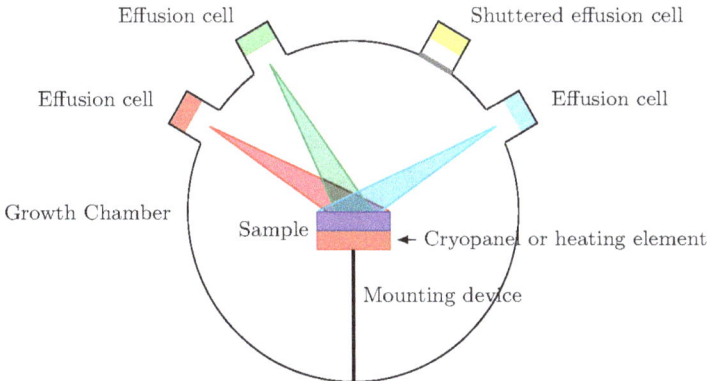

Fig. 2.11   The schematic diagram of the setup for the MBE method. The substrate is just above the holder.

2.2.3.1 *Growth chamber*

The growth chamber of the MBE setup defines an ultra-high vacuum (10^{-8} Pa) region. The chamber houses all of the other parts including the effusion cells, the mounting device and characterization equipment, such as an electron gun and screen used for the reflection high-energy electron diffraction (RHEED) measurements discussed later in section 2.3.1.5.

2.2.3.2 *Effusion cells*

The atoms or molecules that are deposited on the target are provided by Knudsen effusion cells. If a solid source is used, the cells are heated and atoms or molecules sublimate and travel into the growth chamber. A thermocouple is embedded in the cells to carefully monitor the temperature.

It is also possible to install computer controlled shutters in front of each cell to control the thickness of a layer in the film. If multiple effusion cells are used, the shutters are also used to alternate or combine materials in the growth chamber.

MBE is a promising method to carefully grow thin-film HMs on semiconducting substrates. Akinaga *et al.* (2000) grew zinc blende CrAs thin films on top of a GaAs substrate with a magnetic moment of $3 \mu_B$, in agreement with theoretical calculations. Recently, Aldous *et al.* had success epitaxially growing TMEs on semiconductors (Aldous *et al.*, 2012a,b,c).

2.3 Characterization

The characterization of material properties can occur at any step during or after growth. Before growth, it may be necessary to characterize the quality of a substrate or ion source material. Also, the orientation of a crystal or substrate in the growth chamber is important so particles are deposited on the correct surface. During delicate growth procedures, it may also be important to track the growth rate. Once a material is grown, it is necessary to determine the final structure. A number of methods are detailed below that utilize x-rays or electron beams to characterize the composition, structure or orientation of materials.

Spintronic materials will ultimately be utilized in devices that are electronic in nature. Three important aspects of electronic transport properties of spintronic materials are examined. They are the resistivity, the Hall effect, and the anomalous Hall effect.

A number of methods have been developed to characterize the magnetic properties of materials. Two such methods, the Andreev reflection method and the superconducting quantum interference device (SQUID) utilize superconductivity to operate. We discuss how the unique properties of superconductors help characterize the magnetic moment and potential half-metallicity of spintronic materials. Also, we examine how light interacts with the magnetic materials, through the Kerr effect, to give us details about the surface magnetization or the spin polarization of individual atoms in a sample.

2.3.1 *Structural properties*

To determine the structure of bulk materials and surfaces, we are looking for information about the separation between, and the species of, the atoms. In section 1.2.2, we discussed that x-rays are the best choice to probe this length scale without causing further complications interpreting the data. Electron beams are also commonly used in many of the methods described below.

2.3.1.1 *X-ray diffraction*

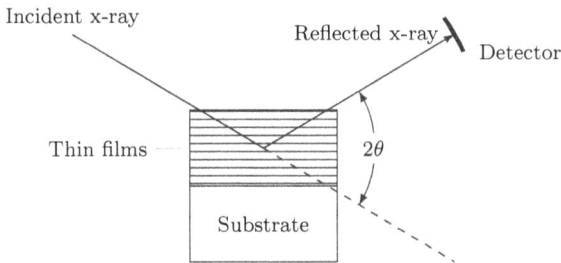

Fig. 2.12   X-ray diffraction.

Figure 2.12 shows an experimental setup of x-ray diffraction. X-rays incident on a crystal structure will scatter due to the electrons in the array. The intensity of the diffracted x-rays are detected by photographic film or digital camera. The intensity of the x-ray peaks are recorded at angles of 2θ from the incident beam to determine the structure.

2.3.1.2   *X-ray photoelectron spectroscopy*

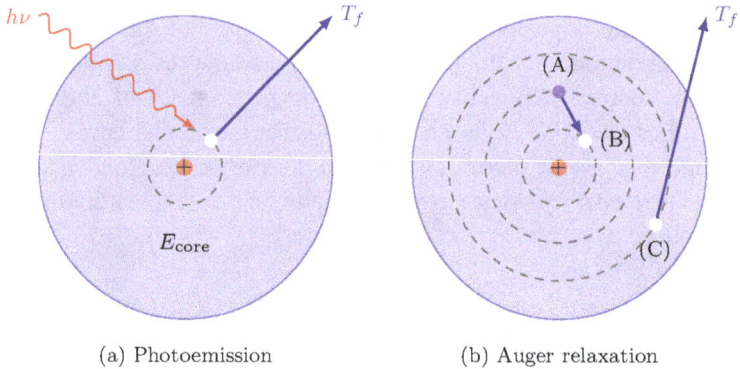

(a) Photoemission   (b) Auger relaxation

Fig. 2.13   (a) Simple illustration of the photoelectric effect on core states. An x-ray with energy $h\nu$ interacts with a core state with energy E_{core} and ejects it from the host atom. The photoelectron leaves the surface with kinetic energy T_f. (b) Picture of the Auger relaxation process. An electron at (A) gives up energy to fill the empty core state at (B). The energy loss causes the ejection of another electron at (C).

X-ray photoelectron spectroscopy (XPS) is capable of surface elemental analysis. By shining monochromatic x-ray photons with energy $h\nu$ on the surface of a material, core electrons with energy E_{core} from atoms near the surface are emitted by the photoelectric effect with final kinetic energy T_f. The processes is shown in Fig. 2.13(a). Since XPS relies on ejection of core-level electrons from atoms it is unable to detect hydrogen or helium atoms in a sample. From conservation of energy, the binding energy of a core electron is

$$E_{bind} = h\nu - T_f. \tag{2.245}$$

The energy measured by the spectrometer is adjusted by the work function $T_{measured} = T_f - \phi$ where ϕ is the work function of the spectrometer. The predominant spectral line in the Si core states is from the 2p core states ($L_{2,3}$) with binding energy around 99 eV.

2.3.1.3   *Auger electron spectroscopy (AES)*

Auger electron spectroscopy (AES) involves the detection of secondary electrons emitted as a result of core hole excitation and relaxation processes (Chang, 1971). Like XPS, the process involves core levels depicted in

Fig. 2.13(b), so it is insensitive to hydrogen and helium atoms. The results are typically used to infer information on the chemical environment of particular atoms in a surface.

The Auger relaxation occurs after an electron in a core state is ejected by means of x-ray photoemission. For Si, a hole in the $L_{(2,3)}$ core state is filled by an outer electron. The energy is either released as a photon or given to another outer shell electron. If the energy given to the outer electron is large enough, this electron can be ejected.

2.3.1.4 *The low energy electron diffraction (LEED) method*

The low-energy electron diffraction (LEED) method uses a low-energy electron beam reflected from the target to display the diffraction pattern on a luminescent screen. The electron beam energy ranges from around 20 to 200 eV. The diffracted beam encounters two or three meshes before hitting the luminescent screen. The first mesh encountered (grid 2 in the figure) is grounded and the remaining meshes accelerate the electron beams. These meshes suppress the secondary electrons that were ejected from the surface.

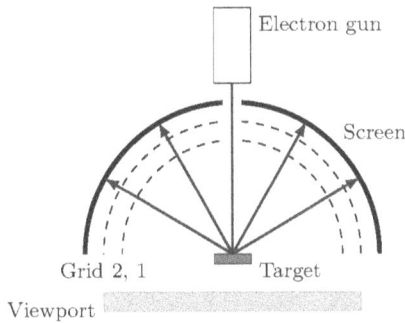

Fig. 2.14 Schematic setup of the LEED measurement. The electron beam hits the target and reflects to the screen behind the viewport. The dashed lines are the meshes.

2.3.1.5 *Reflection high-energy electron diffraction method*

The RHEED method is used to identify and monitor the growth of thin films on a substrate. A schematic setup is shown in Fig 2.15. The electron beam hits the target at a very low angle and is diffracted from the surface layers to a detector on the right. The diffraction pattern is used to identify the surface structure and orientation of the sample.

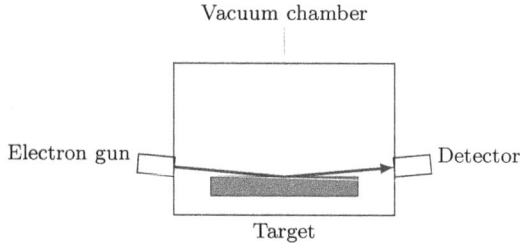

Vacuum chamber

Electron gun

Detector

Target

Fig. 2.15　(a) Schematic diagram for the setup of a RHEED measurement.

While this method may be used on its own, the RHEED method is typically used in conjunction with the MBE growth method to monitor the layer-by-layer growth patterns. As material is deposited on the substrate, the adsorbed atoms diffract the electron beam at different angles, diminishing the intensity of the central peak. The intensity of the central peak oscillates and is maximum when a full monolayer has formed. The process of one layer growth is shown in Fig. 2.16. At (a) $t = 0$, a smooth substrate is shown diffracting the electron beam towards the detector giving a maximum intensity. As particles, shown in red, begin to deposit on the surface in (b), the intensity diminishes. When the surface begins to form a complete layer, as seen in (c), the intensity increases again until in (d) a complete monolayer forms giving a maxima in the intensity. The central spot on the diffraction pattern will oscillate in this manner.

2.3.1.6　*Rutherford backscattering (RBS)*

Rutherford backscattering (RBS) is used to probe the composition of particles near the surface of materials. In this method, α-particles are accelerated towards the surface of interest and the ions are backscattered and their energy measured. The basic idea of RBS is depicted in Fig. 2.17. The energy of the incident particles is of the order of 1000 eV to probe atoms located about 20 nm below the surface. When the incident ion collides with atoms at the surface, the ion and the atom recoil leaving the ion with a much smaller kinetic energy. The energy absorbed by the surface through phonons and intraband transitions is much smaller than the incident energy so it is ignored. The kinetic energy decrease of the reflected particles corresponds to the relative mass of the atom in the surface and the recoil

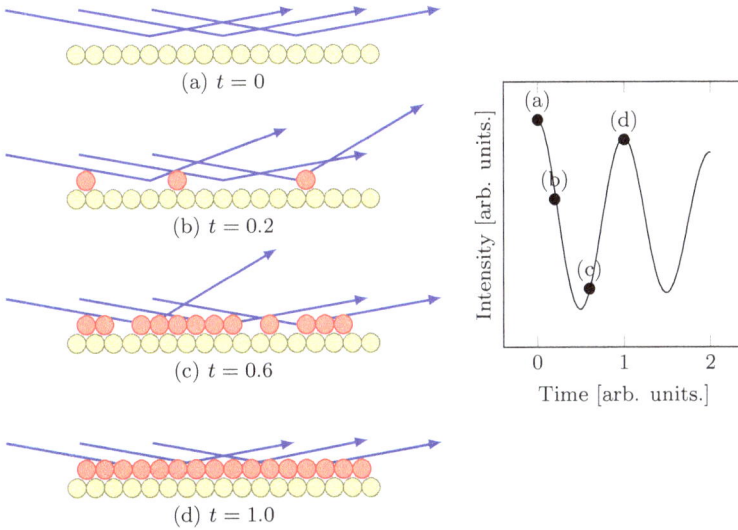

Fig. 2.16   Left: schematic of layer growth as particles (red) are deposited on the surface (yellow). The electron beam of the RHEED diffract off of the atoms. Right: Intensity of RHEED central spot over time.

angle of the ion, thus, by using classical kinematics and energy conservation, it is possible to determine the mass of the atoms near the surface. In Fig. 2.17, beam (a) reflects with large kinetic energy off of a larger atom at the surface. For (c), the incident beam reflects with less kinetic energy off of a light atom on the surface . For (b), the beam reflects from atoms deeper into the material and exits with very low kinetic energy.

2.3.1.7   *Secondary ion mass spectrometry (SIMS)*

The primary idea for the secondary ion mass spectrometry (SIMS) is that the incident ions can eject atoms near the top layer of the films so they can be collected and analyzed. The schematic diagram is shown in Fig. 2.18. By analyzing the mass of the emitted atoms, it is possible to determine the atomic species in the film.

2.3.1.8   *High-resolution transmission electron microscope*

The high-resolution transmission electron microscopy (HRTEM) uses electrons to probe the atomic structure of materials and provide a 2d-projected view of the crystal structure. The magnification of HRTEM can be 10^6

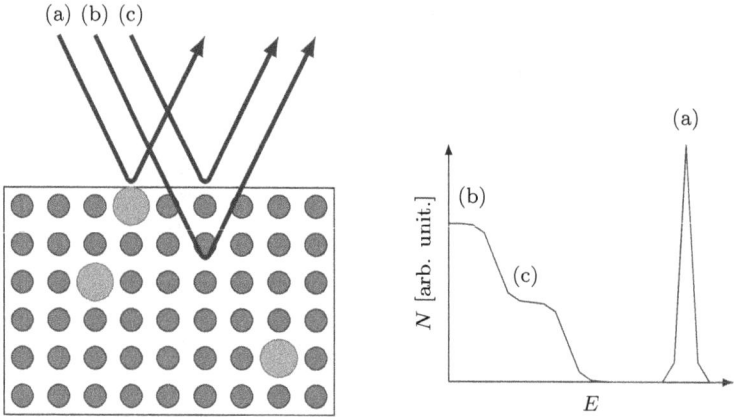

Fig. 2.17   Schematic of the processes for RBS. In the left figure, the arrows (a), (b) and (c) are the incident and the reflected particles that reflect off of the different atoms in the material. The right figure shows the reflected particle count as a function of kinetic energy.

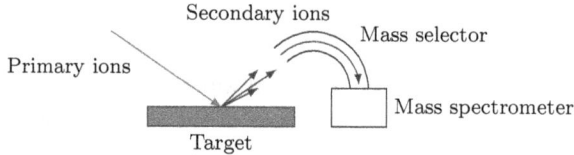

Fig. 2.18   Schematic diagram showing the static and dynamic SIMS.

and the resolution can be as small as 0.5 Å (Kisielowski *et al.*, 2008). A schematic of the microscope is shown in Fig. 2.19.

2.3.2   *Transport*

In modern Si-based devices, the primary operational mechanism is the transport of charge carriers due to a driving voltage V. The purpose of designing Si-based spintronic devices is to manipulate the current technological paradigms to leverage the spin and magnetic properties to do useful work. Therefore, we need to characterize the transport of spin, in addition to charge, carriers in spintronic materials. This section will concentrate on two areas of transport. The first is the measurement of resistivity as is normally studied with conventional electronics. The second is the anomalous

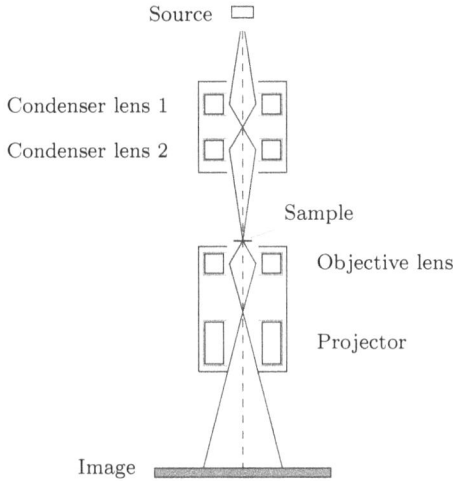

Fig. 2.19   Schematic diagram showing HRTEM.

Hall effect (AHE) in which a current is generated transverse to the applied electric field by the ferromagnetic properties of the material.

2.3.2.1   *Resistivity*

When measuring the resistivity of a bulk sample, such as a rod of length L and cross-sectional area A, a straightforward way is to apply a voltage ΔV across the length L and measure the current I flowing through the sample. The resistivity is

$$\rho = \frac{\Delta V}{I} \frac{A}{L}. \tag{2.246}$$

For thin films and devices on the surface of a substrate, the resistivity is most commonly measured along the surface. The method described above will no longer work because the cross-sectional area is not well-defined. Instead, the standard way to measure resistivity of a surface is by using four probes, shown in Fig. 2.20. The outer probes provide a current through the sample surface while the inner probes measure the voltage drop. When the distance from the probe to the surface boundaries are large and the probe spacing d is small compared to the size of the sample, the resistivity of a bulk sample is

$$\rho = \frac{\pi}{\ln 2} \frac{\Delta V}{I}. \tag{2.247}$$

For different geometries, the numerical factor before the $\Delta V/I$ is different. In general, the resistivity is

$$\rho = C\frac{\Delta V}{I} \qquad (2.248)$$

where the correction factor C is determined by the geometry of the sample. One of the advantages of the four-probe arrangement is to reduce the contact resistance between the sample and the electrode.

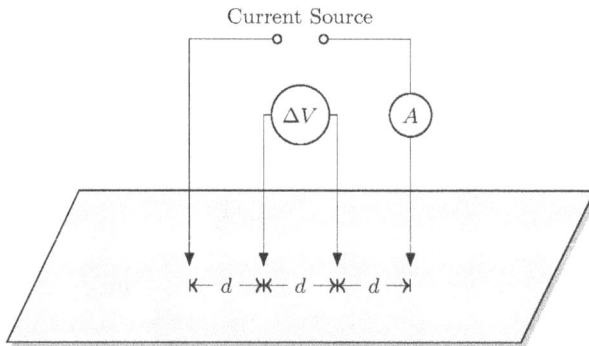

Fig. 2.20   A typical four-probe arrangement (Pesavento *et al.*, 2004)

2.3.2.2   *The Hall effect*

In Fig. 2.21, the schematic diagram of a setup measuring the Hall effect is given. The device is called a Hall bar. A voltage V_a is applied to the bar causing electrons to flow j_x in the $-x$ direction. A magnetic field B_z is also applied to the Hall bar in the direction perpendicular to the electric field. A voltage V_H is measured to characterize the Hall effect. The physical processes are as follows: An electron flows in the $-x$ direction under the applied potential. The electron experiences the Lorentz force due to the applied magnetic field B_z and moves toward the top edge of the slab. The accumulated electrons cause more negative charges at the top edge and missing electrons at the bottom edge leave a positive charge. A voltage, V_H, is created by the accumulation of charges on opposite sides of the bar. After V_H is established, electron flowing from $+x$ to $-x$ direction no longer experiences any deflecting force. This is the Hall effect. The Hall conductivity is the measured current divided by the Hall potential

$$(\sigma_{\mathrm{H}})_{xy} = \frac{I_x}{V_{\mathrm{H}}}. \tag{2.249}$$

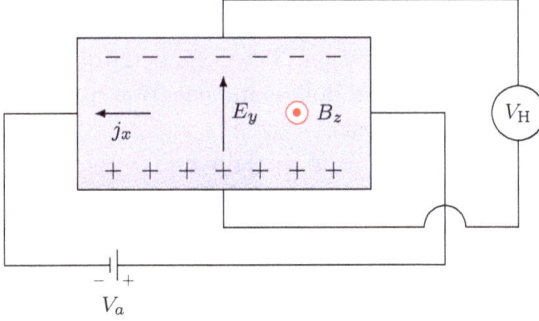

Fig. 2.21 The schematic diagram of a measurement setup of the Hall effect. There is an electric field provided by the voltage V_a and a magnetic field B_z applied to the sample, the blue slab has length L, thickness t and width w. V_{H} is the measured Hall voltage.

2.3.2.3 *Anomalous Hall effect*

The AHE is an additional contribution to the Hall conductivity that happens in ferromagnetic materials, however, the contribution is not due to the additional magnetization of the ferromagnetic material. There are three main contributions to the AHE (Nagaosa *et al.*, 2010) summarized below:

- Intrinsic: The intrinsic AHE is present in ferromagnets even without impurities or disorder. It is due to the $\boldsymbol{k}$-dependence of the periodic part of the Bloch wave function and the Berry-phase curvature

$$\boldsymbol{b}_n(\boldsymbol{k}) = \nabla \times \left\langle u_{n\boldsymbol{k}} \left| \frac{\partial}{\partial \boldsymbol{k}} \right| u_{n\boldsymbol{k}} \right\rangle. \tag{2.250}$$

  The curvature $\boldsymbol{b}_n$ is directly related to the Hall conductivity

$$(\sigma_H)_{xy}^{\mathrm{intrinsic}} = -\frac{e^2}{\hbar} \sum_n \int \frac{d\boldsymbol{k}}{(2\pi)^3} f(\epsilon_n(\boldsymbol{k})) b_n^z(\boldsymbol{k}) \tag{2.251}$$

  where f is the occupation of the state $\epsilon_n(\boldsymbol{k})$ and b_n^z is the z-component of the Berry-phase curvature.

- Skew scattering: the skew scattering mechanism is due to an imbalance of electron band transitions caused by SO coupling near an impurity site. Using perturbation theory with the SO interaction, Smit (1958) found that the transition probability between different momentum states was proportional to the cross product of the momenta

$$|\langle \boldsymbol{k}'|H_{\text{so}}|\boldsymbol{k}\rangle|^2 \sim (\boldsymbol{k}' \times \boldsymbol{k}) \cdot \boldsymbol{M} \qquad (2.252)$$

where $\boldsymbol{M}$ is the magnetization. The asymmetry of the transition probabilities, inserted into the Boltzmann equation, produces a transverse contribution to the current.
- Side-jump mechanism: this is partially due to the transverse displacement of the charge carrier due to the SO interaction with an impurity having a potential $V(r)$

$$H_{\text{so}} = \frac{1}{2m^2c^2}\left(\frac{1}{r}\frac{\partial V}{\partial r}\right)S_z L_z. \qquad (2.253)$$

However, it is often defined as the difference between the true Hall conductivity and the two previous mechanisms

$$(\sigma_H)_{xy}^{\text{side-jump}} = (\sigma_H)_{xy} - (\sigma_H)_{xy}^{\text{skew}} - (\sigma_H)_{xy}^{\text{intrinsic}} \qquad (2.254)$$

so there is no consistently defined mechanism for the side-jump contribution.

2.3.3 *Magnetic characterization*

The important magnetic properties for a spintronic device are the saturation magnetization, magnetic anisotropy, spin polarization at E_F, P, and the Curie temperature T_C. Some of the best and most popular methods to determine the saturation magnetization and P require devices that utilize the unique properties of superconductivity so a brief introduction to the relevant mechanisms in superconductivity are also reviewed in this section.

2.3.3.1 *The saturation magnetization*

For ferromagnetic materials, the relationship between the magnetization M and the applied magnetic field H is described by a hysteresis loop, shown in Fig. 2.22. The loop is determined by measuring $\boldsymbol{B} = \boldsymbol{H} + \boldsymbol{M}$, the magnetic induction, as the ferromagnetic sample is subjected to a range of applied fields $\boldsymbol{H}$.

The most fundamental quantity of ferromagnetic materials is the saturation magnetization M_s. For a demagnetized ferromagnetic sample, the

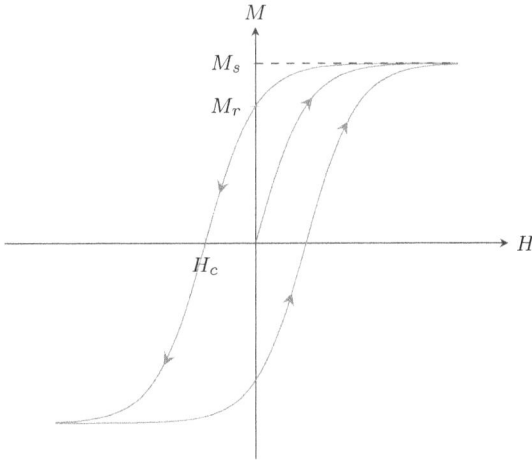

Fig. 2.22   Hysteresis curve.

magnetic domains in the sample are randomly aligned and the net magnetization is $M = 0$. When the ferromagnetic material is subjected to an applied magnetic field H, the magnetic domains in the sample will align in the direction of the field. As the magnitude of H is increased, the domains will align further until all of the moments are pointing in the direction of the applied field. The saturation magnetization M_s is the largest value of M with all moments within the sample are aligned with H. After the magnetic field is reduced to $H = 0$, some magnetic domains remain aligned and the magnetization is the residual magnetization M_r. When the applied field is applied in the opposite direction of the magnetization, domains begin to align in the direction of the opposite field. For some value of the applied field, called the coercivity H_c, the magnetization is driven to zero before reversing direction.

For spintronic materials, it is important that the saturation magnetization is measured at or above RT so the material can be useful in practical devices. There are at least two methods to measure a hysteresis loop: the SQUID based magnetometer and the magneto-optical Kerr effect (MOKE). The SQUID relies on the properties of superconductors, so we will discuss the mechanisms of superconductivity first. It is necessary to start with the discussion of a Cooper pair (Cooper, 1956; Bardeen *et al.*, 1957), the bound state of two electrons through a phonon, and then the Josephson junction (Josephson, 1962).

2.3.3.2   *Cooper pairs*

The microscopic physics of the superconductor are described by Bardeen–Cooper–Schrieffer (BCS) theory but we will only focus on one aspect of the theory: the Cooper pair. A Cooper pair is a bound state of two electrons near E_F interacting through a weak attractive potential (Cooper, 1956). Even though the Coulomb interaction is repulsive, the other electrons screen the repulsion and the interaction of the screened electrons with the lattice results in a net attractive interaction at very low temperatures.

In Fig. 2.23, we provide a simple illustration of a Cooper pair. An electron moving to the right with momentum $\boldsymbol{k}$ (small red circle on the left) induces the motion of a positive ion (larger blue circle on the left) toward left with a speed v_x. The motion of this + ion with velocity $\boldsymbol{v}$ causes a lattice vibration depicted by the wave form at the bottom of the figure. The crest of the wave indicates the largest ionic displacement to the right with respect to its equilibrium position, while the trough of the wave depicts the maximum of displacement of a + ion toward left from its equilibrium position. If the phase relations between the motions of the + ions and another electron exhibit coherence, that is, an electron shown as small filled circle with momentum $-\boldsymbol{k}$ (left moving) can move close to an + ion (large blue circle at right) moving with velocity $\boldsymbol{v}$, then there is a net attractive interaction between the two electrons. They bind together to form a Cooper pair. The binding energy is 2Δ, where Δ is the energy of the bound electron. The superconducting gap is on the order of 0.1 to 1.0 meV.

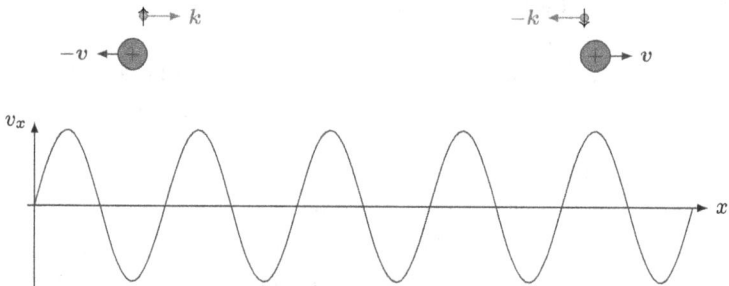

Fig. 2.23   Schematic of a Cooper pair.

The requirements for the electrons forming the Cooper pair are (i) the electrons have opposite momenta, $\boldsymbol{k}$ and $-\boldsymbol{k}$ when no external field is applied; and (ii) their spins are oppositely oriented. The first requirement can be understood from the point of view of FT of the Coulomb interaction between two electrons. This FT is inversely proportional to the square of the absolute difference of their momenta. This means that the repulsion of antiparallel momenta $1/|\boldsymbol{k}-(-\boldsymbol{k})|^2$ is the weakest. The second requirement is from the Pauli principle. In order to experience the coherence between the electrons and the vibrational motion of the ions, it is beneficial to have the two electrons to be close to each other. Due to the Pauli exclusion principle, the electrons in the singlet $S = 0$ state can come close and form a bound pair. It is possible to describe all of the Cooper pairs in a system by a single condensed state with the wave function

$$\Psi_{\text{sc}}(\boldsymbol{r},t) = \sqrt{n(\boldsymbol{r},t)}e^{\mathrm{i}\phi(\boldsymbol{r},t)} \tag{2.255}$$

where n is the density of Cooper pairs and ϕ is the phase of the Cooper pair. The phase ϕ is arbitrary, so it must be gauge invariant. The gauge transformations in the presence of a scalar potential V and vector potential $\boldsymbol{A}$ are

$$V(\boldsymbol{r},t) = V'(\boldsymbol{r},t) + \frac{\phi_0}{2\pi}\dot{\phi}(\boldsymbol{r},t) \tag{2.256}$$

$$\boldsymbol{A}(\boldsymbol{r},t) = \boldsymbol{A}'(\boldsymbol{r},t) + \frac{\phi_0}{2\pi}\boldsymbol{\nabla}\phi(\boldsymbol{r},t) \tag{2.257}$$

where $\phi_0 = 2\pi\hbar/q = h/q$ is defined as the quanta of magnetic flux. For Cooper pairs, there are two electrons, so $q = 2e$.

The bound states of Cooper pairs do not easily scatter from impurities in the metal, since there is insufficient energy to break the pairs. Also, if one electron scatters $\boldsymbol{k} \rightarrow \boldsymbol{k} + \boldsymbol{\Delta k}$, then the other electron will change its momentum to compensate $-\boldsymbol{k} \rightarrow -\boldsymbol{k} - \boldsymbol{\Delta k}$. Since the paired electrons do not scatter, when an external electric field is applied, the cooper pairs experience no resistance and form a supercurrent.

2.3.3.3 *The Josephson junction*

The Josephson junction (JJ) is a pair of superconductors joined by an insulating layer between. A simple diagram of a JJ is presented in Fig. 2.24 where the two superconductors are separated by a thin insulating layer. They can be made of same metals or two different superconductors. There are four effects demonstrated by the JJ. They are

(1) the DC Josephson of two different superconductors effect where supercurrent flows without an applied bias voltage $V = 0$,

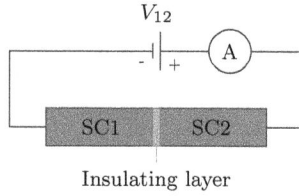

Fig. 2.24   A Josephson junction. The two superconductors can be composed by either identical materials or different superconductors. The region in between the two superconductors is an insulating region.

(2) the AC Josephson effect where alternating supercurrent flows when a small bias voltage $V < 2\Delta/e$ is applied across the junction,

(3) conventional (quasiparticle) current flows only when $V > 2\Delta/e$, and

(4) the supercurrent is very sensitive to the presence of a magnetic field.

The first two effects may be described using a simple tunneling picture. The quasiparticle (QP) current is blocked from flowing at low bias voltage due to the gap in the DOS. We will next describe how the DOS produces current at higher voltages. While a JJ is very sensitive to the magnetic field, we will only demonstrate the sensitivity of the supercurrent to the magnetic field by discussing the application of a SQUID. A SQUID is formed by combining two JJ and is one of the most popular methods for sensitively determining the magnetic properties of a material.

Both the DC and AC Josephson effects are easily described using a simple tunneling picture. The Cooper pair in one superconductor Ψ_1 has a probability of tunneling into the other superconductor and contributing to the supercurrent. Assuming the two superconductors are the same material so the situation is simpler, the probability of tunneling from 1 to 2 is the same as tunneling from 2 to 1. We now represent the tunneling by a constant potential V_{12}, so the time-dependent Schrödinger equations are

$$i\hbar\frac{\partial\Psi_1}{\partial t} = U_1\Psi_1 + V_{12}\Psi_2 \tag{2.258}$$

$$i\hbar\frac{\partial\Psi_2}{\partial t} = U_2\Psi_2 + V_{12}\Psi_1, \tag{2.259}$$

where the superconducting states in the respective superconductors are

$$\Psi_1(t) = \sqrt{n_1(t)}e^{i\phi_1(t)}, \tag{2.260}$$

$$\Psi_2(t) = \sqrt{n_2(t)}e^{i\phi_2(t)}, \tag{2.261}$$

and U_i is the potential of each superconductor. By substituting the wave functions into the Schrödinger's equations and separating the real and imaginary terms, the equations of motion of Cooper pairs across the junction are

$$-\hbar\dot{\delta} = U_2 - U_1, \tag{2.262}$$

$$\frac{\hbar}{2}(\dot{n}_1 - \dot{n}_2) = 2V_{12}\sqrt{n_1 n_2}\sin\delta, \tag{2.263}$$

where $\delta = \phi_2 - \phi_1$ is the phase difference between the superconductors. For superconductors made of different materials, the phases differ even in zero external electric field. The Cooper pairs flow from one superconductor to the other, such that $I_s = \dot{n}_1 = -\dot{n}_2$ is the supercurrent. The phase difference δ between the two superconductors now determines the value of the superconducting current

$$I_s = I_0 \sin(\delta) \tag{2.264}$$

where $I_0 = 2V_{12}\sqrt{n_1 n_2}/\hbar$ is the maximum value of the current. Eq. 2.264 shows that even with zero bias voltage applied across the JJ, supercurrent flows through the junction. This is the DC Josephson effect. The AC Josephson effect occurs when a constant bias voltage V_0 is applied. If the superconductors are the same material, we can set $\delta = 0$ at $t = 0$, the phase δ changes linearly according to Eq. 2.262,

$$\delta(t) = \frac{2qV}{\hbar}t \tag{2.265}$$

$$I_s(t) = I_0 \sin\left(\frac{2eV}{\hbar}t\right) \tag{2.266}$$

Next, we discuss the flow of QP current in the JJ. The superconducting gap 2Δ is the amount of energy required to separate a Cooper pair into two conventional electrons. The gap prevents QPs from entering the conduction band and tunneling across the Josephson barrier at low bias voltage. In Fig. 2.25(a), we show the DOS vs. energy of two identical superconductors near their E_F. The blue regions are the occupied states. The center stripe is the insulating region separating the two superconductors. With no applied voltage bias, the superconducting Cooper pairs lay below the Fermi energy. As the bias voltage increases, the Fermi energy of the right superconductor increases with respect to the left superconductor. Due to the presence of the gap 2Δ, the junction has no states at the Fermi energy until the bias voltage reaches $2\Delta/e$. When the bias voltage exceeds this value, as shown in Fig. 2.25(b), Cooper pairs from the right superconductor will tunnel to the

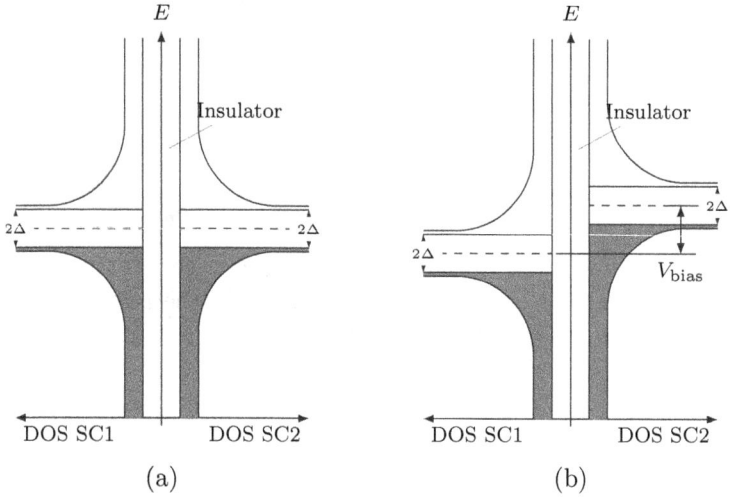

Fig. 2.25   The E vs. the density of states of a junction (a) without any external bias, and (b) with an applied bias.

left superconductor and generate a large QP current due to the large DOS. Before tunneling, the states are below the superconducting gap, so they are Cooper pairs, but after tunneling, the states are above the superconducting gap and the pair must be broken into two quasiparticle charge carriers.

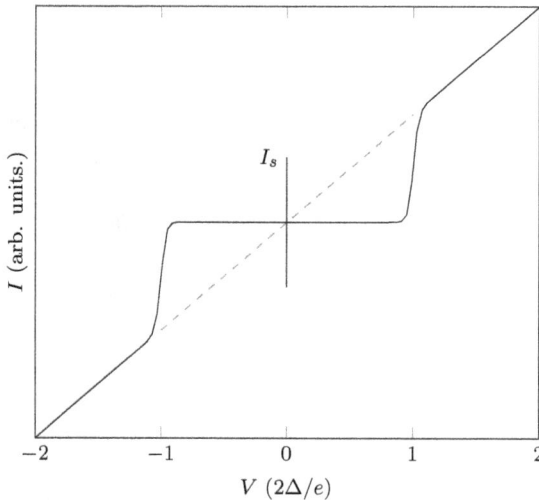

Fig. 2.26   The I-V curve of a JJ formed by identical superconductors.

The *I-V* curve of a JJ including the contribution of the DC Josephson effect is shown in Fig. 2.26. There is no current until the applied voltage reaches $2\Delta/e$. At this value of V, a large jump of the current happens due to the large DOS. After that, the voltage and the current relation is linear as contributed by QPs. The gap in the density of states of the pair of superconductors results in the observed behavior in the JJ.

2.3.3.4   *The SQUID based magnetometer*

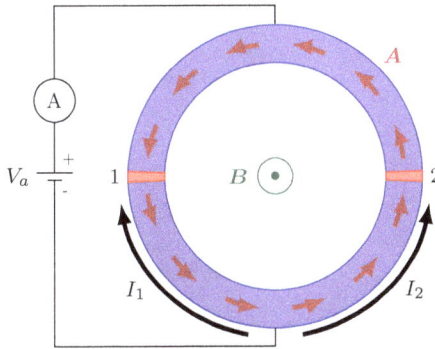

Fig. 2.27   Two JJs in half ring shapes form a SQUID. Tunneling currents in junctions 1 and 2 are sensitive to the presence of a magnetic fluxoid. The current is split between the two paths and tunnel differently due to differing phases across the two junctions. Red arrows indicate the direction of the vector potential inside the half circles.

A SQUID, is formed by two JJ and sensitively detects magnetic flux that pass through the region between the junctions. It is schematically shown in Fig. 2.27. The total current is the sum of the current through the left and right branches

$$I_s = I_0 \sin \delta_1 + I_0 \sin \delta_2 \tag{2.267}$$

$$= 2I_0 \cos\left(\frac{\delta_1 - \delta_2}{2}\right) \sin\left(\frac{\delta_1 + \delta_2}{2}\right) \tag{2.268}$$

$$= 2I_0 \cos\left(\frac{\Delta\phi}{2}\right) \sin\left(\delta_1 - \frac{\Delta\phi}{2}\right) \tag{2.269}$$

where $\Delta\phi$ is the phase difference between the left and right branches. When no magnetic field is applied, the current from the left and right branches combine without interference $\Delta\phi = 0$

$$I_s = 2I_0 \sin(\delta_1).$$ (2.270)

In the presence of a magnetic field, the superconducting phase ϕ is altered due to the vector potential $\boldsymbol{A}$

$$\Delta\phi = \frac{q}{\hbar} \int \boldsymbol{A} \cdot d\boldsymbol{r}$$ (2.271)

$$= \frac{2\pi}{\phi_0} \int \boldsymbol{A} \cdot d\boldsymbol{r}.$$ (2.272)

In the symmetric gauge, the vector potential is

$$\boldsymbol{A} = \frac{B}{2}(-y\hat{\mathbf{x}} + x\hat{\boldsymbol{y}})$$ (2.273)

so the $\boldsymbol{B}$ field passes through the area enclosed by the SQUID. For a closed loop, the integral on the right hand side is the flux passing through the area of the loop

$$\oint \boldsymbol{A} \cdot d\boldsymbol{r} = \Phi_B$$ (2.274)

so the phase difference between the path on the left and right is

$$\Delta\phi = \frac{2\pi\Phi_B}{\phi_0}.$$ (2.275)

The total current is

$$I_s = 2I_0 \cos\left(\frac{\pi\Phi_B}{\phi_0}\right) \sin\left(\delta_1 - \frac{\pi\Phi_B}{\phi_0}\right).$$ (2.276)

In the presence of a constant bias voltage, the sine term oscillates due to the AC Josephson effect, so the maximum current is easily related to the flux through the SQUID by

$$I_{\max} = 2I_0 \left|\cos\left(\frac{\pi\Phi_B}{\phi_0}\right)\right|$$ (2.277)

shown in Fig. 2.28.

A SQUID microscope is a sensitive near-field imaging system for measuring weak magnetic fields by moving a SQUID across an area. The microscope can map out buried current-carrying wires by measuring the magnetic fields as small as $\Phi_B = 10 \times 10^{-7}$ G produced by the currents or can be used to image fields produced by magnetic materials.

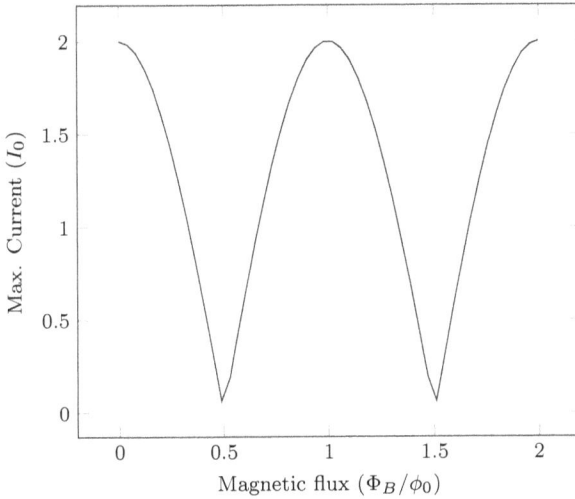

Fig. 2.28   The AC current vs. magnetic flux for a SQUID

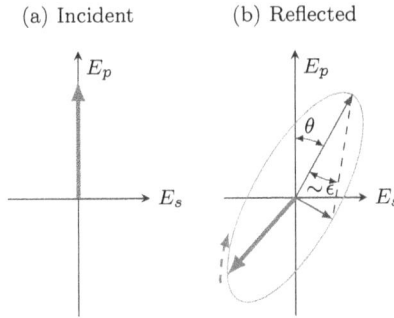

(a) Incident   (b) Reflected

Fig. 2.29   Example of Kerr rotation from (a) the incident p-polarized light phasor to (b) reflected elliptically polarized light phasor. The reflected light has a Kerr rotation angle θ and ellipticity ϵ due to the surface magnetization. The small polarizer angle δ is not shown.

2.3.3.5   *Magneticoptic Kerr effect (MOKE)*

MOKE is based on the Kerr effect: linear polarization of incident light reflected from a magnetized sample is changed to elliptical polarization—the Kerr rotation—shown in Fig. 2.29. The microscopic origin of the Kerr rotation is the SO coupling between the electric field of the incoming light and the spin of the magnetic material (Qiu and Bader, 2000). In the macroscopic picture, the rotation is due to the off-diagonal (anisotropic) elements of the dielectric tensor. MOKE was first applied by Bader *et al.* (1986) to study surface magnetism by measuring the hysteresis loop of a magnetic thin film in the monolayer range. The setup is relatively simple and can be incorporated into the growth vacuum chamber. The schematic diagram is shown in Fig. 2.30.

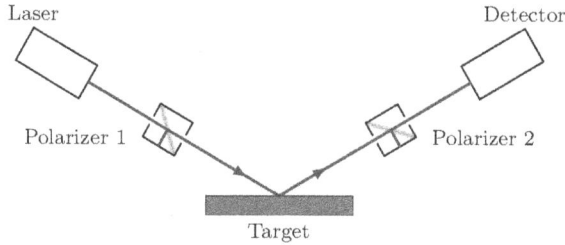

Fig. 2.30 Schematic diagram of the setup of MOKE

Using the first polarizer, the incident light is oriented as either s- (parallel to the plane of incidence) or p- (perpendicular to the plane of incidence) polarized light. The reflected wave will have components E_s and E_p that are parallel and perpendicular to the reflection plane, respectively, with intensity $I = |E_s + E_p|^2$ and complex Kerr angle $\phi = \theta + i\epsilon$. The Kerr angle is proportional to the magnetization of the sample. Usually, the real part is called the Kerr rotation θ and the imaginary part is the Kerr ellipticity ϵ.

The surface magnetization can be oriented by $\boldsymbol{H}$ in the three independent directions relative to the plane of incidence. By adjusting the second polarizer a small angle δ away from the p-axis, the reflected intensity is

$$I = |E_p \sin \delta + E_s \cos \delta|^2 \tag{2.278}$$

$$\approx |E_p \delta + E_s|^2 \tag{2.279}$$

$$= |E_p|^2 |\delta + \phi|^2 \tag{2.280}$$

$$\approx |E_p|^2 (\delta^2 + 2\delta\theta) \tag{2.281}$$

$$= I_0 (1 + 2\theta/\delta) \tag{2.282}$$

where I_0 is the intensity when $\phi = 0$ (the sample is demagnetized and there is no applied field). The reflected light intensity is now linearly dependent on the Kerr rotation and the magnetization of the sample. The Kerr rotation itself does not provide a value of the magnetic moment of the sample.

By adjusting the temperature of the sample, MOKE can also determine the relative temperature dependence of the magnetic saturation $M(T)/M(0)$.

There are two ways to determine the hysteresis loop of a sample. One method is to measure the Kerr rotation as a function of H. The analyzing polarizer is rotated to the maximum intensity to give the Kerr angle. The shape of the loop and the remanence, H_r, are also functions of the layer thickness. The other method is to measure the intensity through the polarizer. With the polarization direction fixed in the polarizer, the intensity of the light passing through the polarizer changes as the strength of the applied magnetic field, H, varies.

2.3.3.6 *Spin polarization*

We will discuss a number of methods used to measure the spin polarization P at E_F in the subsequent sections. The value of P must be measured at or above RT for the device to be useful in spintronic applications. Unfortunately, the method with the best energy resolution, the Andreev reflection method, requires forming Cooper pairs in a superconductor, so this method is unusable at RT. This method has been applied to determine HM properties below RT in CrO_2 (Ji *et al.*, 2001). In the following sections, we discuss the Andreev reflection method, the magnetic tunnel junction (MTJ), and spin-polarized photoemission.

2.3.3.7 *The Andreev reflection method*

The Andreev reflection (AR) method is a tunneling junction between a ferromagnetic metal and a superconductor, such as niobium (Nb). The experimental setup is shown in Fig. 2.31 with the superconductor in the shape of a tip. For this method applied to half-metallic materials, the quasiparticles tunneling from the metal to the superconductor are electrons with only one direction of spin polarization.

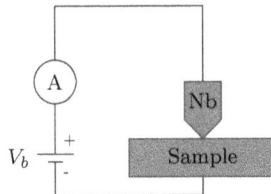

Fig. 2.31 The Schematic diagram for the AR method to detect half-metallic properties. The Nb metal is tapered to serve as a point contact.

Consider the ideal case, at $T = 0\,\mathrm{K}$, E_F of the HM is located at the middle of the superconducting gap. Therefore, there is no way for the QPs with the same spin to form Cooper pairs (up-down pairs) after tunneling to the superconductor side. Consequently, these spin-polarized electrons are reflected from the point of contact back to the HM. There is no tunneling current when the bias voltage is smaller than $\pm\Delta$, where 2Δ is the superconducting gap. The energy diagrams for the tunneling between a NM and a superconductor and between a HM and a superconductor are compared in Fig. 2.32(b) and (c) respectively. Figure 2.32(a) depicts the normal tunneling of both normal metals or HMs when the bias voltage is greater than Δ.

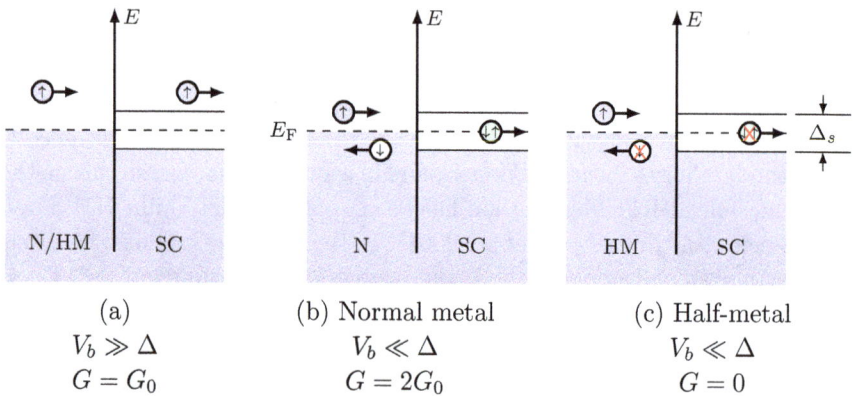

$$
\begin{array}{ccc}
\text{(a)} & \text{(b) Normal metal} & \text{(c) Half-metal} \\
V_b \gg \Delta & V_b \ll \Delta & V_b \ll \Delta \\
G = G_0 & G = 2G_0 & G = 0
\end{array}
$$

Fig. 2.32   Comparison of tunneling from a NM or a HM into a superconductor. (a) When the bias voltage is larger than Δ, both normal metals and HMs conduct. When the bias voltage is smaller than Δ, a normal metal (b) will have enhanced conductivity and a HM (c) will not conduct. The arrows in the circles mark the spin polarizations of the electrons in the metals.

The *I-V* diagrams for the tunneling currents between a normal metal and superconductor and a half metal and a superconductor are shown in Fig. 2.33(a) and (b), respectively. In (a), the current is zero only when the bias voltage is zero. However, there is a bias voltage range of 2Δ where the tunneling current is zero in (b).

2.3.3.8   *Magnetic tunnel junctions (MTJ)*

In section 1.1.3.1, we discussed the GMR device that measured the orientation of a magnetic field based on the measured resistance of a CPP configuration of FM-insultor-FM layers. These devices are also used to de-

(a)

(b)

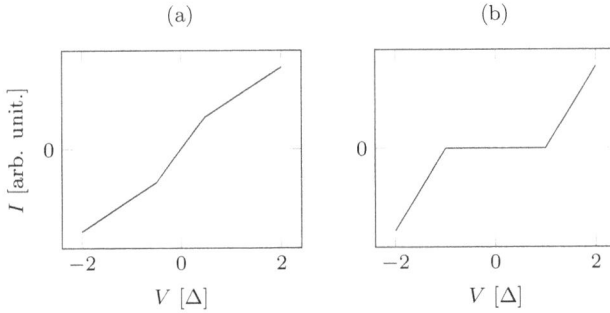

Fig. 2.33 (a) The tunneling I-V curve between a normal metal and a superconductor; (b) the tunneling I-V curve between a HM and a superconductor.

tect spin polarization P in the FM layers. The spin polarizations in such electrodes are defined by the DOS at E_F and are denoted by P_1 and P_2, respectively. These junctions measure the resistance change if the magnetization of the detector is reversed by some controlled magnetic field. For thin-films of FM materials, the schematic setup is shown in Fig. 2.34.

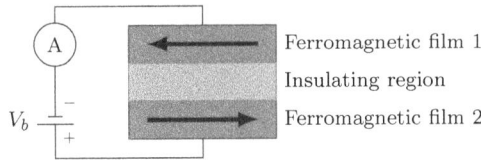

Fig. 2.34 Schematic diagram of a magnetic tunnel junction setup. The first ferromagnetic film acts as a filter of spin polarization while the second film acts as a detector. The black arrows indicate the direction of the magnetization.

The spin polarization of the sample is determined by measuring the tunnel magnetoresistance (TMR). According to the Julliére formula the TMR can be expressed in terms of spin polarizations

$$\text{TMR} = \frac{2P_1 P_2}{1 + P_1 P_2} \qquad (2.283)$$

where P_1 and P_2 are the spin polarizations at E_F in the two ferromagnetic materials, respectively. It is also possible to directly measure the MR by reversing the magnetization in the detecting electrode. The magnetoresistance is

$$\text{MR} = \frac{\rho_{ap} - \rho_p}{\rho_p} \qquad (2.284)$$

where R_p and R_{ap} are the resistivities in the parallel and antiparallel configurations of the magnetic moments. To see how the polarization can be determined, we use the conductance instead of the resistance

$$\text{MR} = \frac{G_p - G_{ap}}{G_{ap}} \tag{2.285}$$

where G_p and G_{ap} are the conductances of the parallel and the antiparallel magnetic configurations, respectively. For the parallel magnetic moment configuration, the majority spin electrons have resulting conductivity proportional to $n_1 n_2$, where n_i ($i = 1$ or 2) are the fraction of majority spins in the DOS of the ferromagnetic layers (1 or 2) at E_F, while the minority spin electrons have $(1 - n_1)(1 - n_2)$. Similarly, when the magnetic moment configuration is antiparallel, the majority spin electron conductivity is related to $n_1(1 - n_2)$ while the minority spin conductivity is $(n_1 - 1)n_2$. The total conductivities for both magnetic moment configurations are

$$G_p = n_1 n_2 + (1 - n_1)(1 - n_2) \tag{2.286}$$
$$G_{ap} = n_1(1 - n_2) + (1 - n_1)n_2. \tag{2.287}$$

G_p and G_{ap} are measured to determine n_1 and n_2. The polarizations in the ferromagnetic layers are

$$P_1 = 2n_1 - 1 \tag{2.288}$$
$$P_2 = 2n_2 - 1. \tag{2.289}$$

2.3.3.9   *Spin polarized photoemission*

Castelliz (1951) measured the magnetic moment of the half-Heusler alloy NiMnSb and found it to be about $4\,\mu_B$, however an integer magnetic moment is not sufficient to determine that the material is half-metallic. To show that a material is a half-metal, an experiment must also determine that one spin channel is metallic while the other is insulating by measuring the spin polarization P.

Photoemission spectroscopy uses the photoelectric effect to measure the binding energies E_bind of electrons in a material. A schematic of the photoemission effect is shown in Fig. 2.35. An incoming photon with energy $h\nu$ and momentum $\hbar\boldsymbol{k}$ can eject an electron in the valence band, initially with energy E_i and momentum $\hbar\boldsymbol{k}_i$ leave it with kinetic energy $T_f = \hbar^2 k_f^2/2m$ and momentum $\hbar\boldsymbol{k}_f$. If the effects of the surface electrons are neglected, the conservation of energy is

$$E_\text{bind} = h\nu - \Phi - T_f \tag{2.290}$$

where Φ is the work function of the material. Similarly, the conservation of momentum parallel to the surface relates the ejection angle of the electron to the initial band momentum $\boldsymbol{k}_i$ such that an ARPES experiment can be preformed. For incoming photons with constant frequency, E_{bind} scales with the final kinetic energy of the ejected electron and the changing position of the detector determines the angle. Finally, a magnet can filter the electrons into up-spin and down-spin components. Figure 2.36 shows a schematic of how the spin-polarized photoelectron intensity of a HM near E_F may appear. One spin channel ($\downarrow$) shows a non-zero intensity at E_F because the DOS is non-zero at E_F so it is metallic. The other spin channel ($\uparrow$) is zero at E_F so it is insulating.

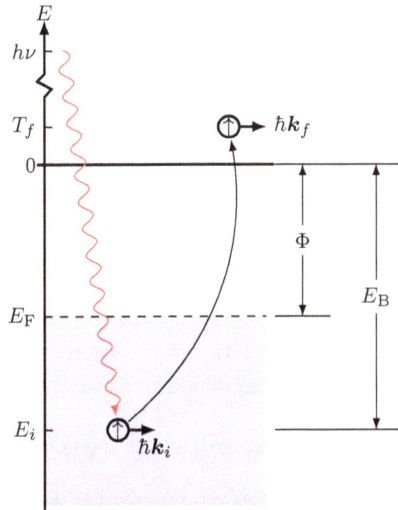

Fig. 2.35   (a) Schematic of the photoemission effect. A photon with energy $h\nu$ ejects an electron from the valence band.

When the binding energy of the ejected electrons is resolved into spin-components, a HM undergoing a photoemission experiment will show a clear contrast between spin channels. Park *et al.* (1998) used spin-polarized photoemission spectroscopy to determine the polarization of the potential HM $La_{0.7}Sr_{0.3}MnO_3$. They showed that one majority spin channel extended up to E_F while the minority spin channel dropped off to zero around $1.0\,\text{eV}$ below E_F. At E_F, the spin polarization was $P = 100\%\pm5\%$. Photoemission is not always an ideal method for determining if a sample exhibits half-metallic properties. Kämper *et al.* (1987) attempted to measure the

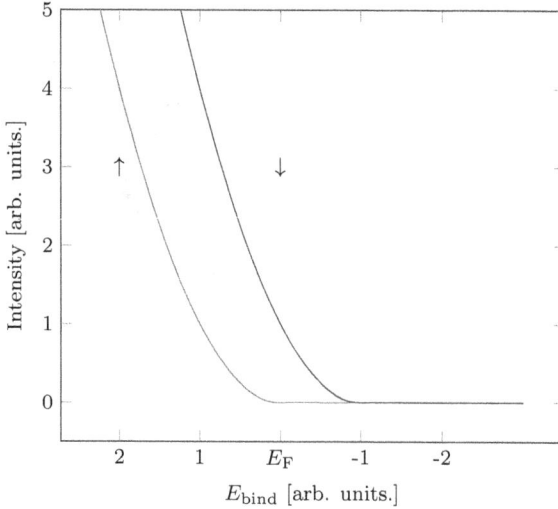

Fig. 2.36   Schematic of the photoelectron intensity of a spin-polarized photoemission experiment.

polarization of CrO_2 and found 100% spin polarization not at E_F, but approximately 2 eV below it. The difficulty in the photoemission experiment may originate in the properties of the sample surface. Kämper *et al.* also found that, while the binding energy was largely unaffected by different surface treatments, the spin polarization drastically changed.

2.3.3.10   *X-ray magnetic circular dichroism (XMCD)*

The magnetic moment of materials with multiple TME species is an aggregate of the individual species moments. X-ray magnetic circular dichroism (XMCD) can be used to determine the local moment of each specie by determining the the absorption of left and right-circular polarized x-rays from core-state photoelectrons (Schütz *et al.*, 1987).

In TMEs, the core states will absorb linearly polarized x-rays in the 100 to 1000 eV range (de Groot, 1994) and transition to the unoccupied d-states above E_F. The absorption rate is proportional to the DOS of the d-states. For circularly polarized x-rays, the absorption process is more complicated. The light will transfer angular momentum $\pm \hbar$ to the photoelectrons and the L_2 and L_3 levels have levels $(l-s)$ and $(l+s)$, respectively, due to SO coupling, so the amount of absorption will be different for spin-↑ and spin-↓ core states. The XMCD signal is the normalized difference in absorption between the right- and left-hand polarized x-rays. It contains the information of exchange splitting and the SO coupling of the initial and the final states. By controlling the energy of the x-ray, it is possible to have the XMCD probe a particular element. The measurements of the photoelectrons with respect to the spin polarization of the sample provide information of the magnetic moment of each TME in the sample.

Chapter 3

Progress in Si-based Spintronics

Making use of the methods discussed in the previous chapter, several research groups around the world have investigated the feasibility of Si-based spintronic materials in device applications. In this chapter, we follow some of the recent developments of Si-based spintronic materials so that these materials can be integrated into more established Si technologies.

Early research was focused on dilutely doping Mn in Si. With the Mn occupying the substitutional (S) site in Si, we explain an anomalous experimental result. Next, we describe a δ-layer of Mn doped in Si with the Mn at an S site (Qian *et al.*, 2006). There is also an open region in crystalline Si, the interstitial (I) site, that a TME can occupy. The issue then is what position the dopant will take. Wu *et al.* (2007) studied a model of doping Mn at the I site in Si and Shauhgnessy *et al.* (2010) compared the energetics of doping Mn and Fe in S and I sites. Yang *et al.* (2013) reported HMs with two layers of Mn atoms occupying I sites in Si and a hole doping between, called trilayers. These results suggest that magnetically doped Si can show half-metallic behavior. Finally, we discuss the developments of high Mn doping concentrations that form MnSi clusters. It is important to understand how these clusters form because they can appear during the growth of Si-based spintronic materials and greatly reduce the large magnetic moment in these materials.

3.1 Dilute doped Mn in Si

It is intuitively clear that magnetic Si samples should be accompanied by doping magnetic TMEs in Si much like previous studies of doping Mn in GaAs, $Mn_xGa_{1-x}As$ (Ohno *et al.*, 1996; Matsukura *et al.*, 1998). Early experiments (Zhou *et al.*, 2007; Bolduc *et al.*, 2005) focused on dilutely doping Mn in Si, Mn_xSi_{1-x} with $x < 0.1\,\%$. Bolduc *et al.* (2005) considered $x = 0.1$ at. % and reported the measured magnetic moment to be $5\,\mu_B/\text{Mn}$: the same as atomic Mn. How can the magnetic moment of Mn doped in Si be as large as the atomic value? The answer depends on the site that the Mn dopant occupies. In this section, we discusses the anomalous experimental findings and the theoretical model used to explain the large magnetic moment.

3.1.1   *Early experiments*

An early experiment of doping Mn into Si was done by Zhang *et al.* (2004). They grew crystalline $Mn_{0.05}Si_{0.95}$ thin film on Si (001) using the vacuum deposition method followed by post-crystallization processing. The 5% concentration of Mn was determined by scanning electron microscope (SEM). XRD was used to determine single phased thin films and the doping causing expansion of the lattice. A SQUID was used to probe the magnetic properties. The single phased film with thermal annealing shows *M-H* hysteresis from 4.2 to 300 K. Standard four-probe scheme used to measure the resistivity as a function of temperature confirms the single phased film showing semiconducting behavior. The Curie temperature is 400 K. No magnetization was detected when the sample is in an amorphous phase.

Later, Bolduc *et al.* (2005) doped Mn in p-type single crystal Si by using the ion implantation growth method. The concentration of holes is $10 \times 10^{19}\,\text{cm}^{-3}$. The pressure and temperature during the doping is $5.0 \times 10^{-6}\,\text{Torr}$ and 350 °C. At this temperature, the grown sample does not show any amorphization. The kinetic energy of the Mn ions is 300 keV. The dosage is between 10^{-5} to $10^{-6}\,\text{cm}^{-2}$. After the implantation, the sample is annealed at 800 °C rapidly for 5 min under N_2 gas. SQUID was used to measure the magnetization. Saturation magnetization for annealed samples at 0.1 at. % shows $5\,\mu_B/\text{Mn}$. On the other hand, the magnetic moment per Mn of annealed sample at 0.8 at. % is $1.5\,\mu_B/\text{Mn}$. The Curie temperature for the sample of 0.8 at. % is 400 K.

It is known that the magnetic moment of a free Mn atom is $5\,\mu_B$ because the two s electrons pair their spins. According to the ionic-model, and based on the fact that Mn substitutionally replaces Ga in GaAs (Ohno *et al.*, 1996), Mn should have a magnetic moment of $3\,\mu_B$ instead of $5\,\mu_B$ at $x = 0.1\,\%$. In the tetrahedral environment, the four neighboring Si each take away one electron from Mn. Three electrons remain on the Mn and align their spin to give the moment. How can a Mn atom surrounded by

Si exhibit the same moment as a free atom?

3.1.2 *Substitutional Mn in Si supercells*

To resolve this discrepancy, Shauhgnessy *et al.* (2009) started with a series of smaller Si supercells by stacking up conventional cells of Si along three directions and replacing one Si with Mn located close to the center of the supercell. Models with 8, 64, and 216 atoms give m less than $5\,\mu_B$ and greater than $3\,\mu_B$ and not necessarily integers because there are interactions between the center Mn and Mn in the neighboring supercells. In order to identify the interaction between neighboring supercells, they used the maximum component of the force acting on the Si atom furthest from the Mn atom. A small force indicates that the Mn is isolated and has a small effect on Si atoms that are far from the Mn. In the 216 atom supercell, the maximum component of the force was $0.08\,\mathrm{eV/\mathring{A}}$.

Next, Shauhgnessy *et al.* tried a 512 atom supercell giving a doping concentration of $0.19\,\mathrm{at.\,\%}$, close to twice the experimental value of $0.08\,\mathrm{at.\,\%}$. The 512 atom supercell has little interaction between neighboring Mn atoms, with the maximum component of the force $0.015\,\mathrm{eV/\mathring{A}}$ even before relaxation, so it gives a moment of exactly $3\,\mu_B$ as expected from S site doping.

The question still remains: how does this type of material get a moment of $5\,\mu_B$ per Mn atom? An alternate set of supercells, where the substituted Si atom is left in the supercell and falls into an I site, explains the physical origin of the unexpectedly large magnetic moment.

3.1.2.1 *The Mn-nn Si-sn Si complex*

During the process of ion implantation, a Si atom is replaced by a Mn atom by breaking the four bonds around the Si atom. However, the Si atom may not leave the crystal. Shauhgnessy *et al.* (2009) modeled this situation and found that the Si atom moves to the I site, which is second neighbor (sn) to its original position.

The alternate supercell, shown in Fig. 3.1, contains a sn Si to the Mn atom and gives a magnetic moment of $5\,\mu_B$. A nearest neighbor (nn) Si occupies an S site between the Mn and sn Si atoms. The three atoms form a complex called the Mn-nn Si and sn Si complex, shown in Fig. 3.2. From the force comparison, the doped Mn located at the center of the supercell remains isolated from the ones in the neighboring image cells. The optimized lattice constant is 5.46 Å, which is close to the lattice constant of pure Si: 5.45 Å.

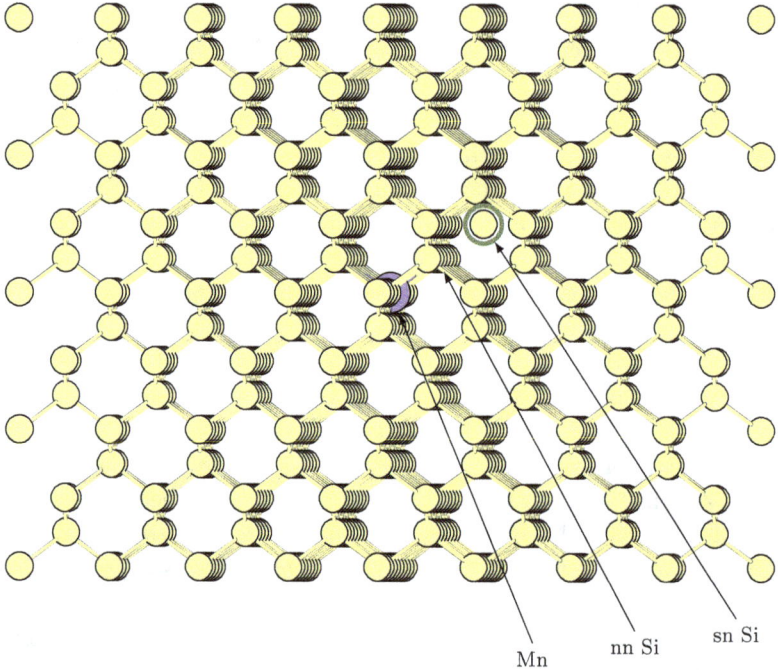

Fig. 3.1   The 513-atom supercell. Mn is shown in blue, Si in yellow, and the sn Si at I site is highlighted in green.

In the 512-atom supercell, the nn Si forms three sp^3 bonds with its neighboring Si atoms and, in Fig. 3.3(a), shows d-p hybridization with the doped Mn. When the replaced Si atom is moved to the I site, shown in Fig. 3.3(b), the sn Si is surrounded by four Si atoms, including the nn Si. This sn Si moves closer to the nn Si because of the difference in the d-p bonding and the sp^3 bonding of its neighbors. The sn Si pulls the sp^3 orbital of the nn Si away from the d-p bond with the Mn. The d-electron of Mn, formerly participating in the d-p hybridization with the nn Si, retreats back to the Mn and increases the local moment at the Mn from 3 to $4\,\mu_B$ according to the first Hund's rule. The electron from the nn Si that was participating in the d-p bond also retreats back to the nn Si. Its spin is now uncorrelated with the retreated d-electron. The spin moment at the Mn polarizes this electron giving a total moment of $5\,\mu_B/\text{Mn}$.

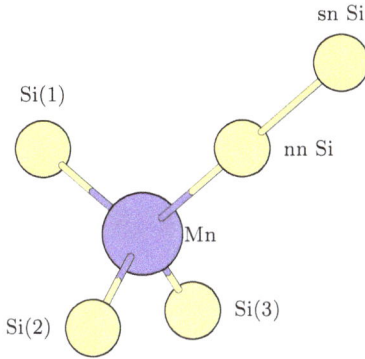

Fig. 3.2   The Mn-nn Si-sn Si complex.

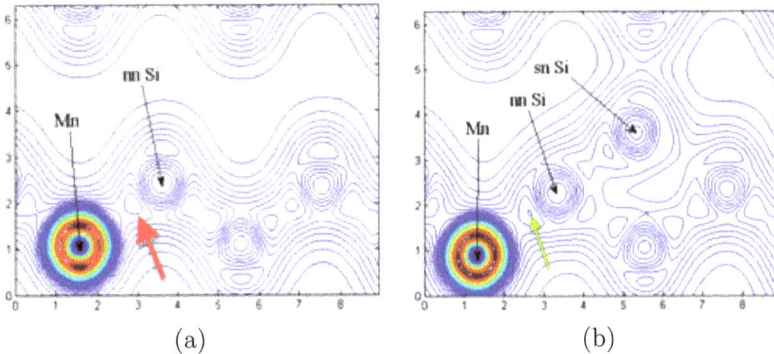

(a)   (b)

Fig. 3.3   The minority spin distributions without (left panel) and with (right panel) the sn Si. Both sections contain the chain of atoms (Mn–Si–Si) along [110] direction of the supercell. The left panel shows the d-p hybridized states of the d_{xy}-type and sp^3. The right panel covers the complex. The green arrow indicates the maximum contour level of the moved sp^3 orbital. This orbital is closer to the nn Si than the maximum in the left panel (the red arrow).

To identify the change in spin polarization due to the introduction of the I site Si, Shauhgnessy *et al.* (2009) plotted the spin polarization density difference between the 513 and 512 atom supercells. Since the atomic positions are slightly different due to the relaxation of the forces, the 512 atom model positions were fixed to match the positions of the 513 atom position, and the contour is shown in Fig. 3.4. The contour levels show the intensity of spin polarization change. Near the Mn site, the four intense lobes identify the d-electrons near the core that increases the local moment from 3 to $4\,\mu_B$. The lower contours in the rest of the supercell represent the polarization of the uncorrelated spins.

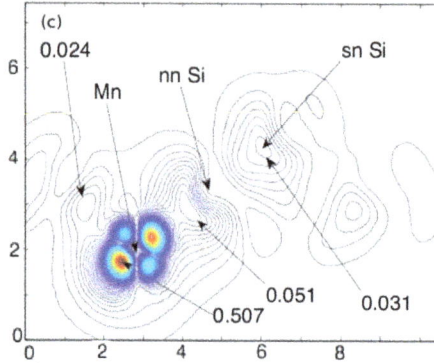

Fig. 3.4　The difference in the spin polarized charge density of the 513 and 512 atom supercells near the complex. The atom positions are fixed to match the relaxed 513-atom supercell. Values of the contour peaks are given in electron/Å^3. All distances are in angstrom. The x-axis of the cross-section is the [110] direction of the supercell and the y-axis coincides with the [001] direction of the supercell.

In summary:

- The S site occupancy of the Mn is consistent with the ion implantation growth method used by Bolduc *et al.* (2005).
- The presence of an sn Si is essential to resolve the discrepancy between the experimental results and the prediction of the ionic model.
- The measured $5.0\,\mu_B$/Mn is contributed by both the local moment at the Mn atom and its polarization effect on the sp^3 orbital of Si.
- This S site occupation is most likely what occurred in the experiments by Bolduc *et al.* (2005).

3.2　Si-based digital ferromagnetic heterostructure

Based on the measured $5.0\,\mu_B$ in Si$_{1-x}$Mn$_x$ with $x = 0.08\%$ by Bolduc *et al.* (2005), Qian *et al.* (2006) designed a δ-layer doped Mn in Si with Mn occupying a S site. This kind of δ-layer is called a digital ferromagnetic heterostructure (DFH) in analogy to the δ-layer doping of Mn in GaAs (Kawakami *et al.*, 2000).

3.2.1 *Perfect δ-layer*

The model is shown in Fig. 3.5(a). The supercell consists of eight conventional cubes in the z-direction. In the x-y plane, the unit cell is defined by the square formed by lattice vectors along the [110] and the [−110] with length $a/\sqrt{2}$, where a is the lattice constant of the conventional cell. It has one Mn in this tetragonal unit cell and the Mn occupies an S site. The optimized lattice constant is 5.45 Å.

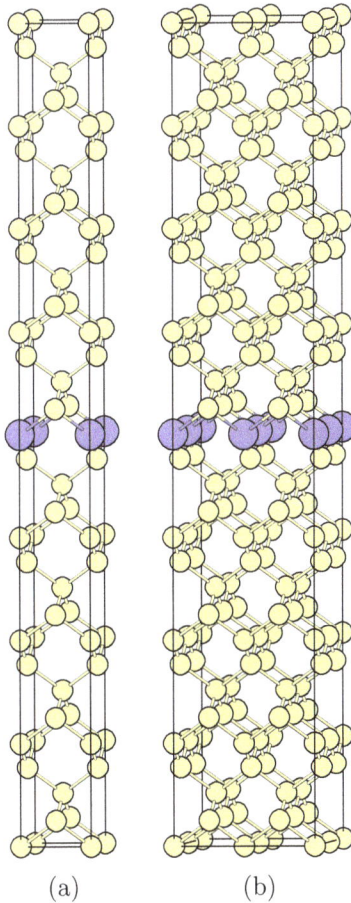

(a)   (b)

Fig. 3.5 DFH models. (a) A 32-atom model created by stacking of 8 conventional Si cells in the z-direction. Si atoms are in yellow, and Mn atoms are in blue. (b) A 64-atom δ-layer model.

Table 3.1 presents the physical properties of the cases without and with the relaxation at this lattice constant. As a result of the relaxation both the DOS of the metallic channel ($\uparrow$) and the gap value in the semiconducting states ($\downarrow$) increases. The DOS is shown in Fig. 3.6. Based on the criteria, this DFH is a HM. In fact, it is a 2-dimensional HM.

Table 3.1   Effect of lattice relaxation due to the presence of the Mn δ-layer in Si on the energies of the ferromagnetic phases (one Mn per unit cell).

Relaxation	DOS$_\uparrow$ (e^-/eV–unit cell)	Moment (μ_B/unit cell)	Total Energy (eV)	Gap (eV)
No	5.06	3	0.0 (Ref.)	0.16
Yes	5.95	3	-0.116	0.25

To see this, we show the band structure and Fermi surfaces of the $\uparrow$ spin channel in Fig. 3.7. Around Γ, there is no band nearby until k is in the middle between Γ and R, the band labeled "1". In (b), this band defines a hole surface (the contour labeled "1"). The labels "2" and "3" around R are filled and specify the electron surfaces. Along the $\Gamma-Z$, there is no intersection of the bands near E_F. One feature of this DFH is the presence of E_F near the top of the VB.

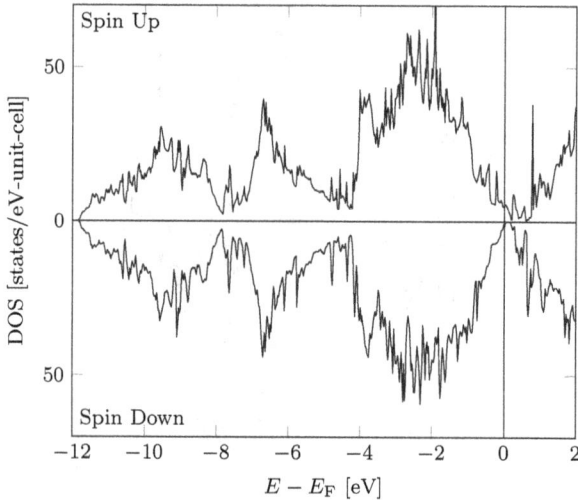

Fig. 3.6   DOS of the DFH showing the half-metallicity. E_F intersects the DOS in the $\uparrow$ spin channel and is in the gap but near the top of the VB of the $\downarrow$ spin states.

Experiments on the well-known half-Heusler (HH) alloy NiMnSb show that its half-metallic properties disappear above 80 K (Hordequin *et al.*, 1996). One possible reason is that thermally excited spin-flip transitions from the metallic channel to the bottom of the conduction band (CB) in the semiconducting channel may be the cause because its E_F lies just below the bottom of the CB (Hordequin *et al.*, 2000). In Fig. 3.9, a schematic energy diagram near E_F for NiMnSb described above and the diagram of the DFH are shown.

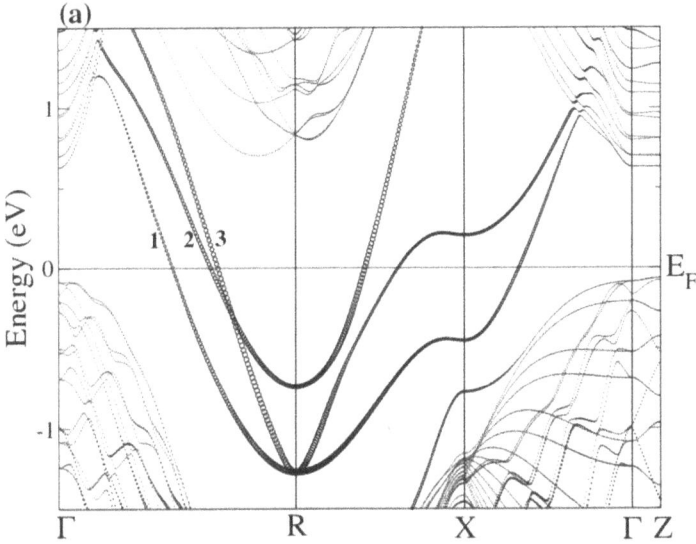

Fig. 3.7 The band structure of the ↑ spin channel.

As argued by Qian *et al.* (2006), the spin-flip transitions are not effective in the Si-based DFH. The E_F is close to the top of the VB. The gap, Δ ($\approx \delta$), is 0.25 eV, which is 10 times of the thermal energy (0.025 eV). Thus, the thermal excitation from E_F to the bottom of the CB is ineffective. On the other hand, spin-flip transitions from the top of the semiconducting VB to the metallic states at E_F are greatly reduced by the matrix element effect: the VB states are localized near Si atoms while the metallic states are formed by the d-states of the Mn, so the overlaps between the initial and the final states of the transitions are small. Fong *et al.* (2010) realized that it is extremely difficult to grow an ideal δ-layer doping experimentally, so they decided to pursue whether imperfections could sustain the half-

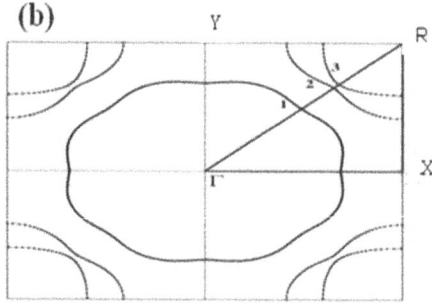

Fig. 3.8   The Fermi surfaces of the ↑ spin channel in the x-y plane of the Brillouin zone.

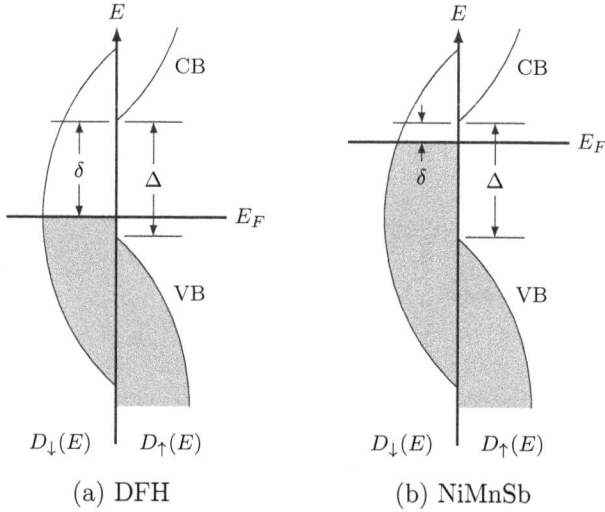

(a) DFH   (b) NiMnSb

Fig. 3.9   Schematic diagrams of the bands near E_F for (a) the DFH and (b) NiMnSb. The energy gap between the E_F and the bottom of the CB is denoted by δ and Δ is the gap of the semiconducting channel.

metallicity. They considered three defects in the δ-layer of the DFH. Two were reported.

3.2.2   Defect models

The model given by Qian *et al.* (2006), containing one Mn in a supercell, cannot be used to examine defects in the δ-layer. Fong *et al.* (2010) expanded the model to contain a total of four Mn atoms in the x-y plane.

The expanded ideal model is given in Fig. 3.5(b). The optimized lattice constant is 5.47 Å for the 64-atom ideal DFH model. Without relaxation, both the 64- and 32-atom models have metallic properties in the ↑ channel. The gaps are 0.21 eV (64-atom) and 0.21 eV (32-atom) in the ↓ channel. For the 64-atom model, the change in total energy after relaxation is 0.141 eV. The ↑ channel is metallic while the ↓ states have a gap of 0.25 eV. All of these agree with the relaxed 32-atom case. Because the 64-atom cell has four Mn atoms, its magnetic moment is 12 μ_B/unit cell. Results of the optimized lattice constants, the gap values in the semiconducting channel and the magnetic moment are summarized in Table 3.2.

Table 3.2 Summary of the optimized lattice constants, the energy gap values in the semiconducting channel, and the magnetic moment of the relaxed 32- and 64-atom supercells.

	Optimized lattice constant (Å)	Energy gap (eV)	Magnetic moment (μ_B/unit cell)
32-atom	5.45	0.25	3
64-atom	5.47	0.25	12

The three models of detects are shown in Fig. 3.10. One Mn in the δ-layer is replaced by a Si in case (a). In case (b), a vacancy is created in the δ-layer. In case (c) an Mn replaces a Si above the δ-layer so the Mn can form a nn pair. This last case simulates the spread of the δ-layer during the growth. The first two cases show the similar half-metallic properties as the ideal case. We plot the DOS of the ideal case and defect case (a) in Fig. 3.11. Both show E_F intersects a finite DOS of the majority-spin channel and is located in the gap of the semiconducting minority-spin channel. The DOS, the gap values and magnetic moments of three defective models are given in Table 3.3.

Table 3.3 Relevant half-metallic properties of the three defect cases.

Case	DOS↑	DOS↓	Gap (eV)	Magnetic Moment (μ_B/unit cell)
1	5.88	0.0	0.29	9.0
2	2.03	0.0	0.49	9.0
3	3.96	3.87	NA	5.52

It is clear that half-metallicity is robust in cases 1 and 2. Case 3 has Mn

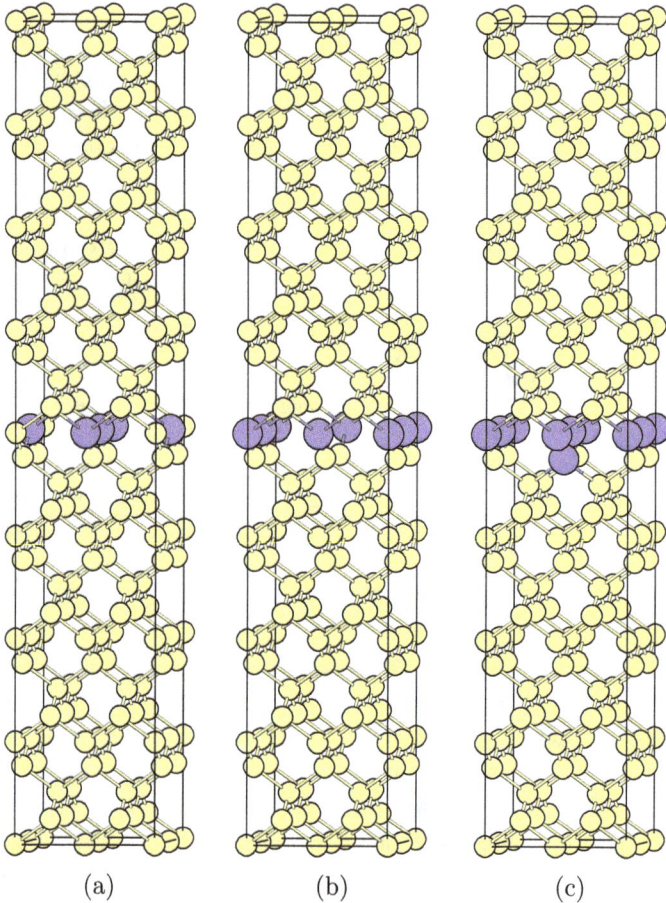

Fig. 3.10   Models of three defect cases: (a) an Mn replaced by a Si, (b) a vacancy where the center Mn is removed, and (c) a Mn over the δ-layer, i.e., the spread of the δ-layer.

atoms in the nn configuration in two adjacent layers. No half-mentallicity is predicted for this case because the d-d interaction between the neighboring Mn reduces the bonding and antibonding gap. During growth, it is necessary to avoid nerighboring Mn-Mn interactions. If the MBE growth method is used, it is necessary to have the beam concentration of Mn to be less than one monolayer. One other possible method can be mentioned: The atomic layer deposition (ALD) method guarantees atomically thick monolayers, but we did not include it in this review because a suitable protocol for ALD of Mn on Si has yet to be established. In summary:

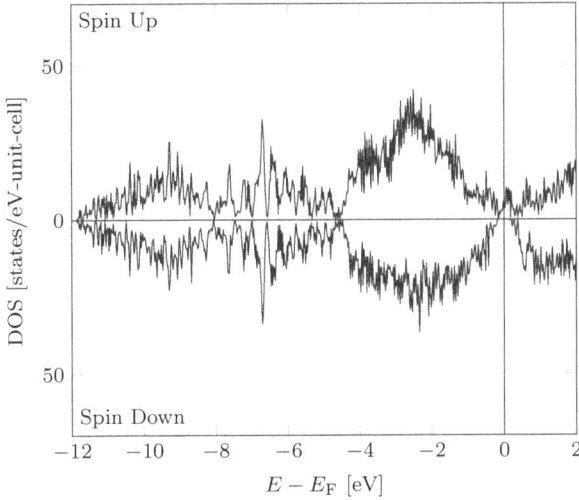

Fig. 3.11   The DOS of defect case 2.

- It is not necessary to have perfect δ-layer doping for a DFH to have half-metallic properties.
- With either 25% vacancy or the presence of 25% Si in the layer, the defective DFH remains half-metallic.
- It is advised to avoid having two Mn atoms form a nn pair.

3.3   Single doping of Fe and Mn in Si

There were a number of developments following the work by Qian *et al.* (2006). Using a similar δ-layer model, Wu *et al.* (2007) used the LAPW method to examine doping Mn at the I site in Si. In Fig. 3.12, pictures of the S and I sites are given. Furthermore, they showed that for 1/4-th monolayer coverage, the I site doped model is a HM. At higher concentrations, the half-metallicity disappears due to the interaction between the interstitial Mn atoms. Zhu *et al.* (2008) suggested that co-doping with As can lower the S site doping barrier of Mn and half-metallicity can be preserved.

To examine the features of doping at the S and I sites, Shauhgnessy *et al.* (2010) investigated the details of the physical origin of the energetic difference between the two doping sites. They tried to answer the following basic questions by examining single dopings of Fe and Mn in Si at the two sites:

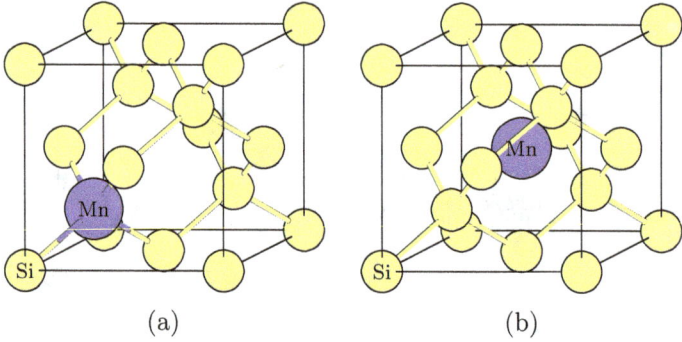

(a)   (b)

Fig. 3.12   Elementary picture of (a) an S site and (b) an I site.

- What are possible distortions caused by the dopings—contraction or expansion? Can they be predicted before growth?
- What are the relative energies between the doping sites, in particular S and a tetrahedral I sites?
- What is the physical origin of the energetic differences?

Table 3.4   Relaxation of the nearest Mn-Si bond length (BL)

TME	Cell Size	Site	Relaxed BL (Å)	Difference from ideal (Å)	Comment
Mn	8	S	2.38	0.02	Expanded
		I	2.41	0.05	
	64	S	2.40	0.04	
		I	2.40	0.04	
	216	S	2.40	0.04	
		I	2.43	0.07	
Fe	8	S	2.32	-0.04	Contracted at S Site
		I	2.36	0.00	
	64	S	2.26	-0.10	
		I	2.40	0.04	
	216	S	2.25	-0.11	
		I	2.40	0.04	

The answer to question 1 is given in Table 3.4. For Mn, the lattice expands independent of the doping sites while Fe doping causes lattice contraction at the S site but lattice expansion at the I site. From atomic radii, 1.26 Å for Fe, 1.35 Å for Mn and 1.32 Å for Si, it is possible to predict the contraction of Fe case at the S site. To answer the second question,

Shauhgnessy *et al.* (2010) started with the expression of the formation energy

$$E_{for} = E_{coh} - N\mu_{Si} - \mu_{TME} \qquad (3.1)$$

where N is the number of Si atoms in a supercell and μ_{Si} is the chemical potential of a Si atom; $-5.433\,eV/Si$. The chemical potential of the TME is denoted by μ_{TME} and E_{coh} is the cohesive energy of the system. They then calculated the difference between the formation energies of the two sites. Since the number of Si atoms in the two cases differs by 1, the difference of the formation energies is

$$\Delta E_{for} = E_{coh}(I) - (E_{coh}(S) + \mu_{Si}) \qquad (3.2)$$

The results are shown in Table 3.5. As the last column of the table shows, the I site has a lower formation energy ($\sim 0.5\,eV$). The charge density plots, Fig. 3.13, in sections containing the atomic chain in the [110] direction of the conventional cell illustrate the microscopic picture of the energetic difference. The different character of the bonding at S and I sites leads to the energetic difference. Fe doping shows similar features.

Table 3.5 The relaxed formation energy difference $\Delta E_{for} = E_{for}(I) - E_{for}(S)$.

Element	Case	ΔE_{form} (eV)
Mn	8-atom	0.164
	64-atom	0.452
	216-atom	0.528
Fe	8-atom	0.415
	64-atom	0.503
	216-atom	0.473

There are several interesting findings when comparing the magnetic moments for various dopings in Table 3.6. The magnetic moments at both sites in the small 8-atom supercell are not integers. This agrees with the result obtained by Wu *et al.* (2007), in which no half metallic properties are found for the I site doping if the layer has more than 1/4 layer coverage. For the Mn case, the moment at the S site is $3\,\mu_B$, which can be characterized by the ionic model. At the I site, the open diamond structure causes electrons of the Mn atom to be attracted by the nuclei of the second neighbor Si atoms. The green arrow in Fig. 3.13 shows one of these electrons. The localized electrons remaining at the Mn contribute to the moment. Due

Table 3.6   Magnetic moments for single doped supercells.

Dopant	Size	Site	Magnetic Moment (μ_B/TME)
Mn	8-atom	S	2.97
		I	3.16
	64-atom	S	3.00
		I	3.00
	216-atom	S	3.00
		I	3.00
Fe	8-atom	S	1.76
		I	2.05
	64-atom	S	0.00
		I	2.00
	216-atom	S	0.00
		I	2.00

to the presence of four Si nuclei, the ionic model also applies and predicts $3\,\mu_B$ in agreement with the calculated value.

Based on the ionic model, the calculated magnetic moment for Fe should be $4\,\mu_B$. Fe is a valence 8 element, so in the presence of four nn Si at the S site, Fe shares four of its electrons to form d-p hybridized bonds and has four remaining electrons to align and produce the moment. However, at the S site, the contractions of the nn Si/Fe bond lengths (shown in Table 3.4) reduces the volume around the Fe preventing its localized electrons from aligning their spin moments and satisfying Hund's first rule. Two localized electrons flip their spins; consequently, no net magnetic moment is produced (as seen in Table 3.6). At the I site, four of its electrons are shifted to the sn Si atoms. The volume resulting from the expansion of the nn Si/Fe bond lengths is not enough to allow all four electrons to align their spins. Instead, only one of the four electrons flips its spin to give the moment of $2\,\mu_B$ (Table 3.6).

In summary:

- At an I site, both Fe and Mn cause local expansion. This explains results on Fe doping in Si showing an increase in measured lattice constant of the alloy with respect to the pure Si (Su *et al.*, 2009; Shaughnessy *et al.*, 2010).
- At an S site, Fe doping shows a contraction while doping Mn shows an expansion. It is possible to predict this behavior from the atomic radii.
- At an I site, there is no bond formation. Due to the open diamond-type

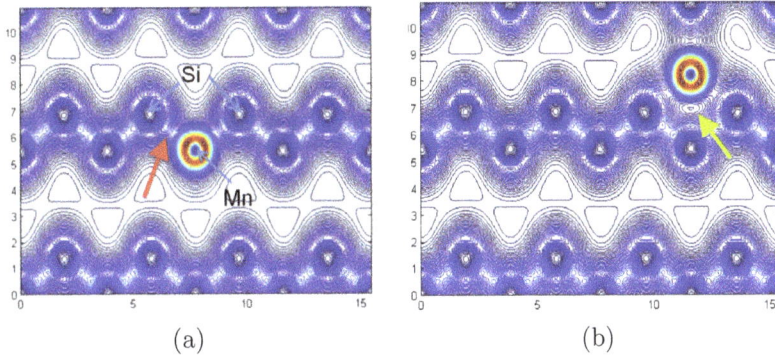

(a)   (b)

Fig. 3.13 Charge density plots. (a) Mn occupies an S site. A d-p bond, indicated by the red arrow, is formed between Si and Mn. (b) There is no bond formation between Si and Mn when Mn occupies an I site. Instead, the Mn shifts its charge to an open region towards its second neighbor, indicated by the green arrow.

structure, the d-electrons of the TME shift their charges to the open regions.

- The physical origin of the energetic differences between the two sites is due to the requirement of breaking four Si-Si bonds at the S site. The energy cost is not compensated by the formation of new d-p hybridized bonds. The energy difference between the two sites is about 0.5 eV.

- The ionic model does not predict the magnetic moment because of the local distortion. The difference is a demonstration of the Pauli principle and the effect of confinement.

3.4 Trilayers

Based on the DFH and the results of examining the doping Mn in either the S or I site, Yang *et al.* (2013) designed trilayers in which Mn occupies I sites. The goal was to explore new Si-based HMs and spintronic materials similar to metallic layered devices. The starting model is provided in Fig. 3.14 with the Mn shown in blue and Si shown in yellow. Any device fabricated from this type of trilayer can operate under the CPP mode.

To operate in the CPP mode, a possible configuration has the current flowing perpendicular to the layer planes. However, the model shown in Fig. 3.14(a) has Si layers between the Mn (blue) layers that would inhibit the current flow. To enhance the current, Yang *et al.* (2013) doped the Si layers between the Mn layers. Either n-type or p-type doping in Si can be

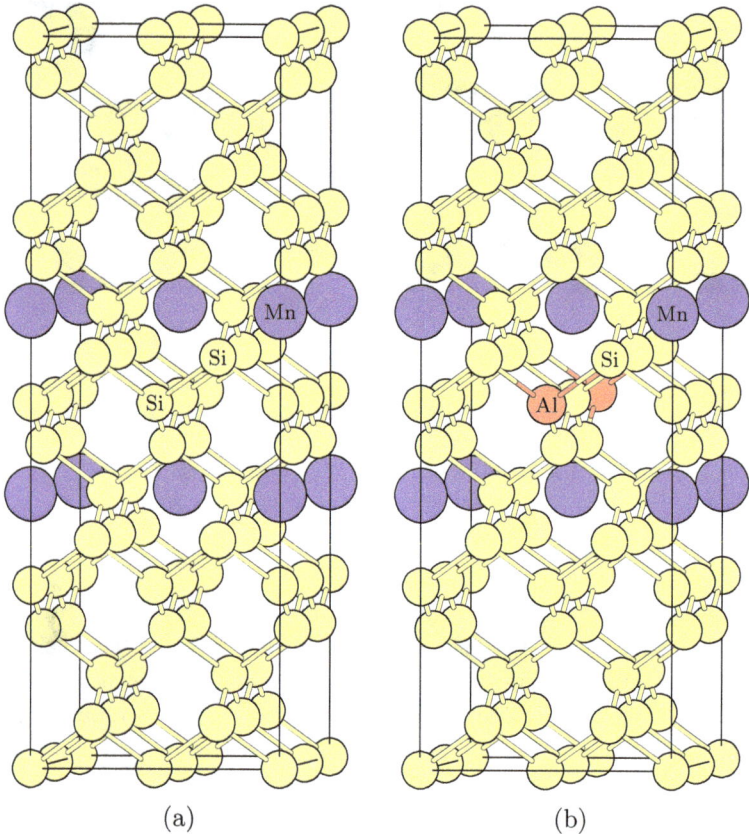

Fig. 3.14 (a) A model of a trilayer having two layers of Mn (blue cirles) at I sites. Si atoms are shown as yellow circles. (b) A trilayer with hole-doping (red circles; either Al or Ga) in the Si layers between the two Mn layers.

easily carried out by experiments because of mature Si technologies. For n-type doping, the donor states may reduce the gap of the Si. A smaller gap of the host material can reduce the likelihood of getting a HM. Furthermore, the donors may overlap their wave functions with the ones of the Mn—consequently, the sample may be a metal. For a metal such as Fe, the spin polarization at E_F is in general less than 100 %. Alternatively, p-type doping can possibly increase the gap. Additionally, the presence of holes causes some of the electrons of the Mn to shift their charge toward the holes. This process may enhance the formation of local moments of the sample— the remaining electrons at the Mn atoms form local moments. Yang *et al.* (2013) considered p-doping with Al and Ga, respectively. The hole-doped model is shown in Fig. 3.14(b) with either Al or Ga in red. The relaxed lattice constants, magnetic moment and the energy gap in semiconducting channel are summarized in Table 3.7. Without hole doping, the trilayer is not a half metal. This result is consistent with the one given by Wu *et al.* (2007)—greater than one forth coverage of the layers results in no half-metallicity. The gaps for Al and Ga dopings are slightly larger than room temperature and were determined by the GGA, which in general underestimates the magnitude of the gap but can be improved by using the *GW* method (Hedin, 1965; Damewood and Fong, 2011).

Table 3.7   Lattice constants in the x-y and z (specified by the c/a ratio) directions, magnetic moment, and energy gaps for the doped trilayers.

Hole dopant	xy-lattice constant (Å)	c/a factor	Magnetic moment (μ_B/unit cell)	Energy gap (eV)
None	5.473	3.970	12.3	NA (ref.)
Al	5.470	3.970	14	0.030↑
Ga	5.458	3.995	14	0.044↑

Since the DOS of doping either Ga or Al exhibit the same qualitative features, we only plot the DOS of the Al doping in Fig. 3.15. E_F is in the gap of the ↑ channel and intersects the finite DOS of the ↓ states. Therefore, with integer magnetic moments, both hole-doped trilayers are HMs. Based on the results of single doping of Mn in Si, the magnetic moment is $3\,\mu_B$/Mn. In this model, there are four Mn atoms. The moment per unit cell should be $12\,\mu_B$/unit cell. However, the calculated moment is $14\,\mu_B$/unit cell. The other $2\,\mu_B$/unit cell come from the dangling bonds of Si atoms which are nn to the Al site. To show this, Yang *et al.* (2013) calculated the difference of the spin polarization density without and with hole-doping. The section that contains Al, Si and Mn is shown Fig. 3.16. The contour labeled (i) has a value of $0.0205\,$electron/Å^3 and is close to where Si atoms are located. This contour and the corresponding one at the other side (ii) are associated with the dangling bonds of the nn Si to the Al. These results are summarized as follows:

- With a full layer of Mn at an I site, half-metallicity can be obtained by doping with either Al or Ga between the Mn layers.
- The dangling bonds, due to the hole-doping, can contribute to the magnetic moment of the sample.

Since this type of doping of Mn does not need to break a bond between

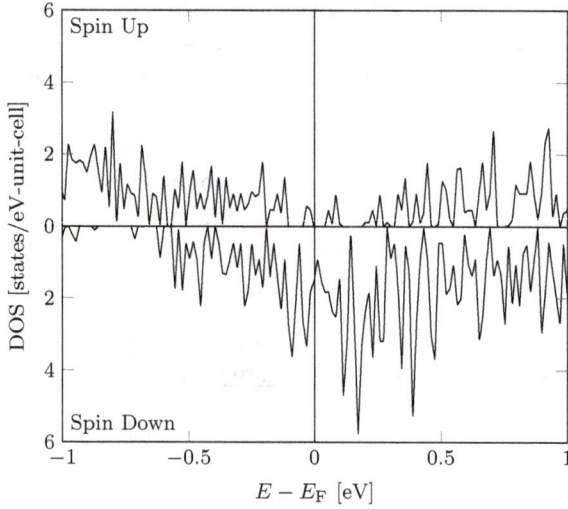

Fig. 3.15   The DOS of the trilayer with additional Al doping. E_F falls within the gap of the ↑ channel, fulfilling the first criteria for a HM.

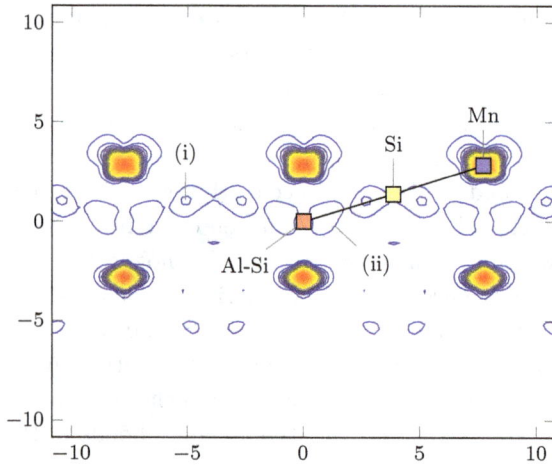

Fig. 3.16   The differences of spin densities with and without hole-doping. The section contains the Al, Si and Mn shown in Fig. 3.14. The label Al-Si indicates the position where Al replaces Si.

pairs of Si, the energy barrier for this doping is at least 0.5 eV/atom less as compared to the doping at an S site. Furthermore, the hole-doping in Si should be feasible due to the established technology. With the gap expected to be larger than RT and E_F located at the top of the valence band, the doped trilayers are promising spintronic materials for device applications.

3.5 MnSi Clusters

In the following section, we discuss the cluster form of MnSi alloys. This form of MnSi is characterized by the grouping, or clustering, of Mn atoms such that each Mn atom has one or more nn Mn. These structures can easily grow at or under the surface of Si-based spintronic materials. Since the Mn-atoms in clusters tend to have at least one nn Mn-Mn bond, half-metallicity will be destroyed and the magnetic moment will reduce. In order to avoid the formation of MnSi clusters in spintronic materials, we need to understand how they grow and what their properties are.

3.5.1 *Growth of MnSi*

3.5.1.1 *Grown by ion implantation*

Among the methods of growth, ion implantation is more popular than other methods because it is possible to focus on the area that is implanted. In general, during the implantation, Mn atoms are believed to (a) form clusters, in particular silicides, and (b) diffuse through channels of I sites at RT (Rueß *et al.*, 2013).

Zhou *et al.* (2007) has carried out ion implantation experiments to dope Mn atoms in Si. The forms include clusters of Mn_4Si_7, $Mn_{11}Si_{14}$, $Mn_{15}Si_{24}$, and $Mn_{27}Si_{47}$. The clusters can be thought of as the stacked tetragonal structure of $MnSi_{1.7}$ having the lattice constant $a = 5.5\,\text{Å}$ and c ranging from 17.5 to 117.9 Å. The Mn-doped samples were grown on p-type (B-doped with concentration of $1.2 \times 10^{-19}\,\text{cm}^{-3}$) Si (100) surface. A Mn^+ ion beam with three fluences ($10^{15}\,\text{cm}^{-2}$, $10^{16}\,\text{cm}^{-2}$ and $5 \times 10^{16}\,\text{cm}^{-2}$) and kinetic energy of 300 keV were incident on the Si surface. Three different concentrations (0.08 at. %, 0.8 at. % and 4 at. %) of Mn-implanted at a depth of $(258 \pm 82)\,\text{nm}$ from the surface were formed, respectively. The samples were held at 350 °C to avoid amorphization and the incident beam was set to form a 7° angle with the normal to prevent channeling. To heal the samples, the rapid thermal annealing (RTA) process was followed right after implantation and the samples were heated to 800 °C for 5 min in N_2 gas environment.

By using HRTEM, nanoparticles were observed in the samples. Their sizes depend on the way they are grown. For fluence of $5 \times 10^{16} \, \mathrm{cm}^{-2}$, the diameter was $20 \, \mathrm{nm}$, for $10^{16} \, \mathrm{cm}^{-2}$, the size was $10 \, \mathrm{nm}$ and for $10^{15} \, \mathrm{cm}^{-2}$, it was $5 \, \mathrm{nm}$. The nanoparticles, however, were not detectable by XRD using conventional sources or synchrotron radiation due to the small number of nanoparticles. Instead, Zhou *et al.* used a grazing incidence XRD method to detect the nanoparticles. This method uses very small angles with respect to the surface to diffract x-rays from only a few layers into the sample. Using an incidence angle of $4°$, diffraction peaks at $2\theta = 42°$ and $46.3°$ were found. Si crystals do not have these diffraction peaks, but the extra peaks could be due to the $MnSi_{1.7}$.

Hysteresis loops were obtained in the $1 \times 10^{16} \, \mathrm{cm}^{-2}$ and $5 \times 10^{16} \, \mathrm{cm}^{-2}$ samples at $10 \, \mathrm{K}$ after RTA. The hysteresis loops of the samples are schematically drawn in Fig. 3.17(a). The $5 \times 10^{16} \, \mathrm{cm}^{-2}$ sample is nearly flat showing very weak magnetic properties. The $1 \times 10^{16} \, \mathrm{cm}^{-2}$ sample exhibits ferromagnetic properties with a saturation magnetic moment $0.21 \, \mu_B/\mathrm{Mn}$ and the coercivity $(275 \pm 25) \, \mathrm{Oe}$.

The origin of the magnetization in these samples is difficult to infer due to the presence of the nanoparticles. To clarify this issue, Zhou *et al.* (2007) cooled the $1 \times 10^{16} \, \mathrm{cm}^{-2}$ sample using either zero-field cooling (ZFC) or field cooling (FC) in a field of $50 \, \mathrm{Oe}$. The magnetization behavior of these two methods may indicate that the magnetization is due to the nanoparticles. The magnetization as a function of temperature is shown in Fig. 3.17(b). The ZFC curve shows the characteristic magnetization behavior superparamagnetism, which is expected for magnetic nanoparticles.

The question can be asked at this point: Is there any evidence of forming clusters using ion implantation to dope Mn into n-type and pure Si substrates? The answer is "yes" for pure Si substrates. Ko *et al.* (2008) reported two samples grown at $350 \, \mathrm{K}$ with fluences of $10^{16} \, \mathrm{cm}^{-2}$ and $2.0 \times 10^{16} \, \mathrm{cm}^{-2}$. Before carrying out any measurement, RTA was performed for $10 \, \mathrm{min}$ at temperature of $900 \, °\mathrm{C}$ for both samples. Fe and Cr have been detected by secondary ion mass spectroscopy (SIMS), but not by either XRD or RBS. No reason was given why the TMEs were not detected using these methods; however, it could be the amounts of Fe and Cr were small or they are deep below the surface. The Mn concentrations near $170 \, \mathrm{nm}$ was $1.92 \times 10^{20} \, \mathrm{cm}^{-3}$ for the low fluence sample and $8.94 \times 10^{20} \, \mathrm{cm}^{-3}$ for the sample with high fluence. HRTEM showed precipitates in both samples. XRD identifies that the precipitates were Mn_4Si_7. The magnetic moment vs. T measurements find a peak at $47 \, \mathrm{K}$, agreeing

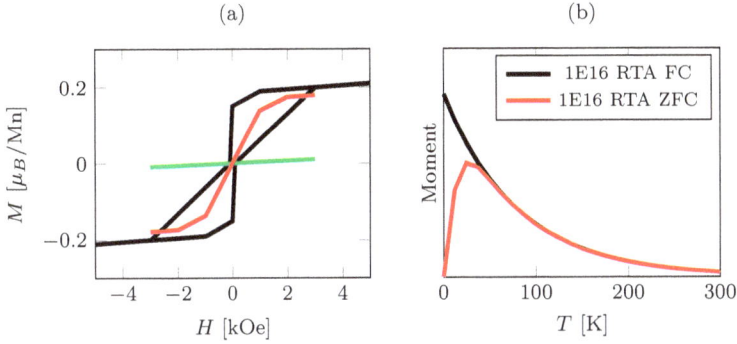

Fig. 3.17 (a) The hyteresis loops of 10^{16} cm^{-2} (black) and 5×10^{16} cm^{-2} (green) at 10 K. Also, the hysteresis loop of 10^{16} cm^{-2} at 100 K (red). (b) Moment vs. temperature for the 10^{16} cm^{-2} fluence sample. The black curve is with field cooling (FC) and the red curve is under zero field cooling (ZFC).

with earlier experiment on Mn$_4$Si$_7$. Therefore, the cluster form of Mn$_4$Si$_7$ is sure to exist when Mn is implanted into Si.

3.5.1.2 *Eptiaxial Mn thin-film growth on Si(111)*

Karhu *et al.* (2010) grew thin films of Mn on Si(111) surface by using two methods: (a) solid phase epitaxy (SPE) and MBE. The film thicknesses are of the order 5 monolayers (ML). The film grown by SPE exhibits serious interfacial roughness. The MBE grown sample is much better in the quality of the interface. MnSi$_7$ clusters are detected in these films. There appears to be lattice mismatch between MnSi (111) and Si(111) of

$$(a_{\mathrm{MnSi}}/a_{\mathrm{Si}}) \cos(30°) - 1 = -3.0\,\% \tag{3.3}$$

which causes an uniaxial anisotropy with the easy axis in the plane. It also influences the measured values T_{C} (45 K). The T_{C} for the bulk MnSi is 29.5 K. The magnetic structures in these films show helical phase, chiral state, in particular for the MBE grown film.

3.5.1.3 *Controlled diffusion of Mn at the surface of Si*

Rueß *et al.* (2013) used a dual chamber with a ultra high vacuum system separating the setup for the scanning microscopy from growth sources. Typical pressure is 10^{-11} to 10^{-10} mbar. High-quality Si(100) with (2x1) reconstruction surface are used as the substrates to deposit Mn atoms at temperatures $-80\,°$C and $50\,°$C to avoid diffusion of Mn atoms. The Mn coverage is about 0.08 ML. Then, the samples are encapsulated by epilayer of Si of 2 nm using a sublimation source of Si and high-substrate temperature, 350 °C. The resultant samples are composed of Mn is δ-layer forms. The full-width-half-maximum of the δ-layer is 1.8 nm. With this thickness, we anticipate that these samples are not HMs because the Mn layers are in the nn configurations. However, it is encouraging that new experimental techniques are developing.

3.5.2 *Theoretical Studies of MnSi$_x$, $x \approx 1.7$*

Miyazaki *et al.* (2008) used the full-potential linearized augmented plane wave (FP-LAPW) method to calculate electronic properties of MnSi$_{1.74}$ using their own experimental information about the structure. The structure is considered as the Nowotny chimney-ladder phase with two basic tetragonal sublattices of Mn and Si, respectively. The unit cell for the sublattices are shown in Fig. 3.18(a) and (b).

In Fig. 3.18(b) the basic tetragonal cell for Mn is given. There are four Mn in this basic cell: one in the corner, two on the faces, and one in the center. Lattice constants are about $a_{Mn} = 5.53$ Å and $c_{Mn} = 4.37$ Å, respectively. The Si basic cell is shown in Fig. 3.18(a). None of the Si atoms are located on the faces or edges of the basic cell. The lattice constants are about $a_{Si} = \sqrt{2} \times 3.91$ Å $= 5.53$ Å and $c_{Si} = 2.51$ Å. To combine the sublattices, it is necessary to stack the unit cells vertically so their respective c lattice constants are similar. Four basic unit cells of Mn ($4c_{Mn}$) and about seven basic unit of Si ($7c_{Si}$) stacked give the form Mn$_4$Si$_7$, stacked cells should be combined as shown in Fig. 3.18(c).

In Fig. 3.18(c), the entire MnSi$_{1.7}$ cell is shown. However, high-resolution neutron diffraction data shows the Si atoms exhibit helical structure which is incommensurable with the unit cell defined by the basic cells of Mn. The incommensurability is indicated by a irrational c-axis ratio between Mn and Si sublattices of $\gamma \approx 1.7361$. This structure is deduced from combining the neutron diffraction data and the use of the modulated crystal structure approach, discussed in section 1.2.4.

The calculation of MnSi$_{1.7}$ is approximated by using 16 Mn and 14 Si in a monoclinic cell, a commensurate structure. The lattice constants of a and c are 5.527 Å and 17.467 Å, respectively. The result shows that the approximated sample is a semiconductor with a gap of about 0.6 eV. However, because the c is about three times larger than a, the shortness of Γ–Z axis in the Brillouin zone causes a difficulty to identify whether the

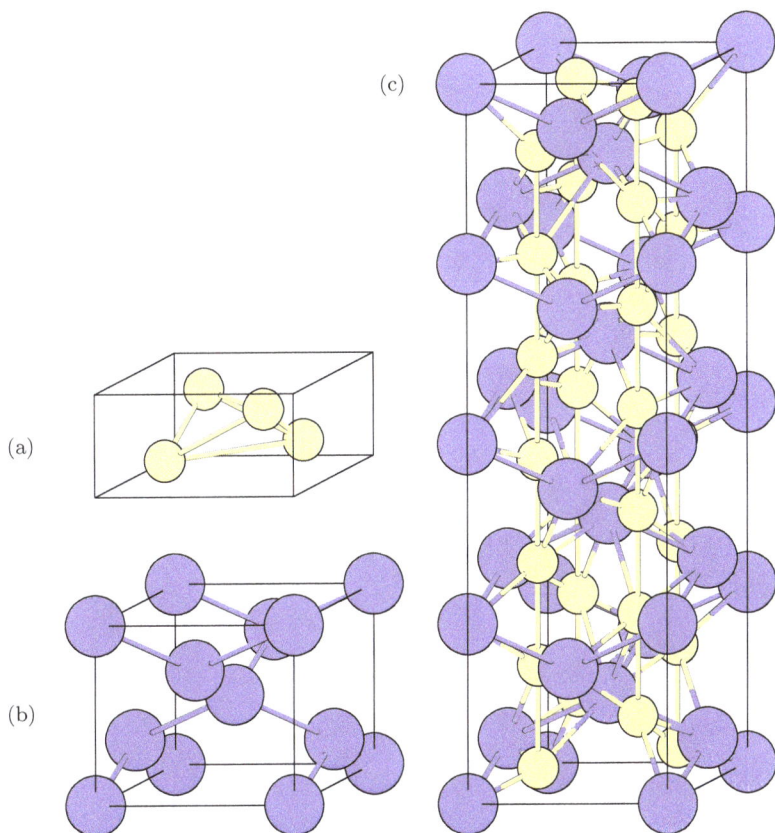

Fig. 3.18 The stacking of Mn and Si basic cell to form an approximation to MnSi$_{1.7}$: (a) the basic Si cell, (b) the basic Mn cell, and (c) the full cell.

sample is a direct or an indirect gap semiconductor. Earlier calculations give gaps between 0.32 eV and 0.70 eV.

Bibliography

Adamo, C. and Barone, V. (1999). Toward reliable density functional methods without adjustable parameters: The PBE0 model, *Journal of Chemical Physics* **110**, 13, p. 6158.

Akinaga, H., Manago, T., and Shirai, M. (2000). Material Design of Half-Metallic Zinc-Blende CrAs and the Synthesis by Molecular-Beam Epitaxy, *Journal of Applied Physics* **39**, 11B, p. L1118.

Aldous, J. D., Burrows, C. W., Maskery, I., Brewer, M., Pickup, D., Walker, M., Mudd, J., Hase, T. P. A., Duffy, J. A., Wilkins, S., Sánchez-Hanke, C., and Bell, G. R. (2012a). Growth and characterisation of NiSb(0001)/GaAs(111)B epitaxial films, *Journal of Crystal Growth* **357**, 15, p. 1.

Aldous, J. D., Burrows, C. W., Maskery, I., Brewer, M. S., Hase, T. P. A., Duffy, J. A., Lees, M. R., Sánchez-Hanke, C., Decoster, T., Theis, W., Quesada, A., Schmid, A. K., and Bell, G. R. (2012b). Depth-dependent magnetism in epitaxial MnSb thin films: effects of surface passivation and cleaning, *Journal of Physics: Condensed Matter* **24**, 14, p. 146002.

Aldous, J. D., Burrows, C. W., Sánchez, A. M., Beanland, R., Maskery, I., Bradley, M. K., dos Santos Dias, M., Staunton, J. B., and Bell, G. R. (2012c). Cubic MnSb: Epitaxial growth of a predicted room temperature half-metal, *Physical Review B* **85**, 6, p. 060403.

Andersen, O. K. (1975). Linear methods in band theory, *Physical Review B* **12**, 8, p. 3060.

Austin, B., Heine, V., and Sham, L. (1962). General Theory of Pseudopotentials, *Physical Review* **127**, 1, p. 276.

Bader, S. D., Moog, E. R., and Grünberg, P. (1986). Magnetic hysteresis of epitaxially-deposited iron in the monolayer range: A Kerr effect experiment in surface magnetism, *Journal of Magnetism and Magnetic Materials* **53**, 4, p. L295.

Baibich, M. N., Broto, J. M., Fert, A., Van Dau Nguyen, F., Petroff, F., Etienne, P., Creuzet, G., Friederich, A., and Chazelas, J. (1988). Giant magnetoresistance of (001) Fe/(001) Cr magnetic superlattices, *Physical Review Letters* **61**, 21, p. 2472.

Bardeen, J., Cooper, L. N., and Schrieffer, J. R. (1957). Theory of Superconductivity, *Physical Review* **108**, 5, p. 1175.

Binasch, G., Grünberg, P., Saurenbach, F., and Zinn, W. (1989). Enhanced magnetoresistance in layered magnetic structures with antiferromagnetic interlayer exchange, *Physical Review B* **39**, 7, p. 4828.

Bloch, F. (1929). Über die Quantenmechanik der Elektronen in Kristallgittern, *Z. Physik* **52**, 7-8, p. 555.

Blöchl, P. E. (1994). Projector augmented-wave method, *Physical Review B* **50**, 24, p. 17953.

Bogolyubov, N. N. and Tyablikov, S. V. (1959). Retarded and Advanced Green Functions in Statistical Physics, *Soviet Physics Doklady* **4**, p. 589.

Bohm, D. and Pines, D. (1951). A Collective Description of Electron Interactions. I. Magnetic Interactions, *Physical Review* **82**, 5, p. 625.

Bolduc, M., Awo-Affouda, C., Stollenwerk, A., Huang, M., Ramos, F., Agnello, G., and LaBella, V. (2005). Above room temperature ferromagnetism in Mn-ion implanted Si, *Physical Review B* **71**, 3, p. 033302.

Borca, C., Komesu, T., Jeong, H.-K., Dowben, P., Ristoiu, D., Hordequin, C., Nozières, J., Pierre, J., Stadler, S., and Idzerda, Y. (2001). Evidence for temperature dependent moments ordering in ferromagnetic NiMnSb(100), *Physical Review B* **64**, 5, p. 052409.

Born, M. and Oppenheimer, R. (1927). Zur Quantentheorie der Molekeln, *Annalen der Physik (Leipzig)* **389**, 20, p. 457.

Bozorth, R. M. (1968) *Ferromagnetism* (Van Nostrand).

Callen, H. (1963). Green Function Theory of Ferromagnetism, *Physical Review* **130**, 3, p. 890.

Castelliz, L. (1951). Eine ferromagnetische Phase im System Nickel-Mangan-Antimon, *Monatshefte fur Chemie* **82**, 6, p. 1059.

Ceperley, D. M. and Alder, B. J. (1980). Ground State of the Electron Gas by a Stochastic Method, *Physical Review Letters* **45**, 7, p. 566.

Chang, C. C. (1971). Auger electron spectroscopy, *Surface Science* **25**, 1, p. 53.

Chikazumi, S. (1964) *Physics of Magnetism* (Wiley).

Cole, L. and Perdew, J. P. (1982). Calculated electron affinities of the elements, *Physical Review A* **25**, 3, p. 1265.

Cooper, L. N. (1956). Bound Electron Pairs in a Degenerate Fermi Gas, *Physical Review* **104**, 4, p. 1189.

Damewood, L. J. (2013). Theoretical Models of Spintronic Materials, Ph.D thesis, University of California, Davis, CA. 3602035.

Damewood, L. J. and Fong, C. Y. (2011). Local field effects in half-metals: A GW study of zincblende CrAs, MnAs, and MnC, *Physical Review B* **83**, 11, p. 113102.

Daughton, J. M. (1992). Magnetoresistive memory technology, *Thin Solid Films* **216**, 1, p. 162.

de Groot, F. M. F. (1994). X-ray absorption and dichroism of transition metals and their compounds, *Journal of Electron Spectroscopy and Related Phenomena* **67**, 4, p. 529.

de Groot, R. A., Mueller, F. M., van Engen, P. G., and Buschow, K. H. J. (1983).

New Class of Materials: Half-Metallic Ferromagnets, *Physical Review Letters* **50**, 25, p. 2024.

De Wolff, P. M. (1974). The Pseudo-Symmetry of Modulated Crystal Structures, *Acta Crystallographica Section A: Crystal Physics Diffraction Theoretical and General Crystallography* **30**, 6, p. 777.

Demidov, E. S., Danilov, Y. A., Podol'skiĭ, V. V., Lesnikov, V. P., Sapozhnikov, M. V., and Suchkov, A. I. (2006). Ferromagnetism in epitaxial germanium and silicon layers supersaturated with managanese and iron impurities, *JETP Letters* **83**, 12, p. 568.

DuBois, D. F. (1959a). Electron interactions, *Annals of Physics* **7**, 2, p. 174.

DuBois, D. F. (1959b). Electron interactions, *Annals of Physics* **8**, 1, p. 24.

Filippi, C., Singh, D. J., and Umrigar, C. J. (1994). All-electron local-density and generalized-gradient calculations of the structural properties of semiconductors, *Physical Review B* **50**, 20, p. 14947.

Fong, C. Y. and Cohen, M. (1970). Energy Band Structure of Copper by the Empirical Pseudopotential Method, *Physical Review Letters* **24**, 7, p. 306.

Fong, C. Y., Shauhgnessy, M., Snow, R., and Yang, L. H. (2010). Theoretical investigations of defects in a Si-based digital ferromagnetic heterostructure – a spintronic material, *Physica Status Solidi C: Current Topics in Solid State Physics* **7**, 3-4, p. 747.

Görling, A. (1996). Exact treatment of exchange in Kohn-Sham band-structure schemes, *Physical Review B* **53**, 11, p. 7024.

Gunnarsson, O., Jonson, M., and Lundqvist, B. I. (1976). Exchange and correlation in atoms, molecules and solids, *Physics Letters A* **59**, 3, p. 177.

Hamann, D. R., Schlüter, M., and Chiang, C. (1979). Norm-Conserving Pseudopotentials, *Physical Review Letters* **43**, 20, p. 1494.

Hedin, L. (1965). New Method for Calculating the One-Particle Green's Function with Application to the Electron-Gas Problem, *Physical Review* **139**, 3A, p. A796.

Herring, C. (1940). A New Method for Calculating Wave Functions in Crystals, *Physical Review* **57**, 12, p. 1169.

Hohenberg, P. and Kohn, W. (1964). Inhomogeneous electron gas, *Physical Review* **136**, 3B, p. B864.

Hordequin, C., Pierre, J., and Currat, R. (1996). Magnetic excitations in the half-metallic NiMnSb ferromagnet: From Heisenberg-type to itinerant behaviour, *Journal of Magnetism and Magnetic Materials* **162**, 1, p. 75.

Hordequin, C., Pierre, J., and Currat, R. (1997). How do magnetic excitations behave in half metallic ferromagnets? The case of NiMnSb, *Physica B: Condensed Matter* **234–236**, p. 605.

Hordequin, C., Ristoiu, D., Ranno, L., and Pierre, J. (2000). On the cross-over from half-metal to normal ferromagnet in NiMnSb, *European Physical Journal B* **16**, 2, p. 287.

Hybertsen, M. S. and Louie, S. G. (1985). First-Principles Theory of Quasiparticles: Calculation of Band Gaps in Semiconductors and Insulators, *Physical Review Letters* **55**, 13, p. 1418.

Hybertsen, M. S. and Louie, S. G. (1986). Electron correlation in semiconductors and insulators: Band gaps and quasiparticle energies, *Physical Review B* **34**, 8, p. 5390.

Jackson, J. D. (1998). *Classical Electrodynamics*, 3rd edn. (Wiley).

Ji, Y., Strijkers, G. J., Yang, F. Y., Chien, C. L., Byers, J. M., Anguelouch, A., Xiao, G., and Gupta, A. (2001). Determination of the Spin Polarization of Half-Metallic CrO_2 by Point Contact Andreev Reflection, *Physical Review Letters* **86**, 24, p. 5585.

Johnson, D. (1974). Local field effects and the dielectric response matrix of insulators: A model, *Physical Review B* **9**, 10, p. 4475.

Josephson, B. D. (1962). Possible new effects in superconductive tunnelling, *Physics Letters* **1**, 7, p. 251.

Julliere, M. (1975). Tunneling between ferromagnetic films, *Physics Letters A* **54**, 3, p. 225.

Kämper, K., Schmitt, W., Güntherodt, G., Gambino, R., and Ruf, R. (1987). CrO_2—A New Half-Metallic Ferromagnet? *Physical Review Letters* **59**, 24, p. 2788.

Karhu, E. A., Kahwaji, S., Monchesky, T. L., Parsons, C., Robertson, M. D., and Maunders, C. (2010). Structure and magnetic properties of MnSi epitaxial thin films, *Physical Review B* **82**, 18, p. 184417.

Kasuya, T. (1956). A Theory of Metallic Ferro- and Antiferromagnetism on Zener's Model, *Progress of Theoretical Physics* **16**, 1, p. 45.

Kawakami, R. K., Johnston-Halperin, E., Chen, L. F., Hanson, M., Guébels, N., Speck, J. S., Gossard, A. C., and Awschalom, D. D. (2000). (Ga,Mn)As as a digital ferromagnetic heterostructure, *Applied Physics Letters* **77**, 15, p. 2379.

Kisielowski, C., Freitag, B., Bischoff, M., van Lin, H., Lazar, S., Knippels, G., Tiemeijer, P., van der Stam, M., von Harrach, S., Stekelenburg, M., Haider, M., Uhlemann, S., Müller, H., Hartel, P., Kabius, B., Miller, D., Petrov, I., Olson, E. A., Donchev, T., Kenik, E. A., Lupini, A. R., Bentley, J., Pennycook, S. J., Anderson, I. M., Minor, A. M., Schmid, A. K., Duden, T., Radmilovic, V., Ramasse, Q. M., Watanabe, M., Erni, R., Stach, E. A., Denes, P., and Dahmen, U. (2008). Detection of Single Atoms and Buried Defects in Three Dimensions by Aberration-Corrected Electron Microscope with 0.5-Å Information Limit, *Microscopy and Microanalysis* **14**, 5, p. 469.

Kleinman, L. and Bylander, D. (1982). Efficacious Form for Model Pseudopotentials, *Physical Review Letters* **48**, 20, p. 1425.

Ko, V., Teo, K. L., Liew, T., and Chong, T. C. (2008). Origins of ferromagnetism in transition-metal doped Si, *Journal of Applied Physics* **104**, 3, p. 033912.

Koelling, D. D. (1972). Linearized form of the APW method, *Journal of Physics and Chemistry of Solids* **33**, 6, p. 1335.

Koelling, D. D. (1975). Use of energy derivative of the radial solution in an augmented plane wave method: application to copper, *Journal of Physics F: Metal Physics* **5**, 11, p. 2041.

Kohn, W. and Sham, L. (1965). Self-consistent equations including exchange and correlation effects, *Physical Review* **140**, 4A, p. A1133.

Kresse, G. and Joubert, D. (1999). From ultrasoft pseudopotentials to the projector augmented-wave method, *Physical Review B* **59**, 3, p. 1758.

Kübler, J. (1984). First principle theory of metallic magnetism, *Physica B+C* **127**, 1-3, p. 257.

Kucheyev, S. O., Williams, J. S., and Pearton, S. J. (2001). Ion implantation into GaN, *Materials Science and Engineering* **33**, 2-3, p. 51.

Langreth, D. C. and Perdew, J. P. (1980). Theory of nonuniform electronic systems. I. Analysis of the gradient approximation and a generalization that works, *Physical Review B* **21**, 12, p. 5469.

Lee, M. J. G. and Falicov, L. M. (1968). The de Haas-van Alphen Effect and the Fermi Surface of Potassium, *Proceedings of the Royal Society of London. Series A, Mathematical and Physical Sciences* **304**, 1478, p. 319.

Levy, M. and Perdew, J. P. (1985). The Constrained Search Formulation of Density Functional Theory, in *Density Functional Methods In Physics* (Springer US, Boston, MA), p. 11.

Lieb, E. H. (1979). A lower bound for Coulomb energies, *Physics Letters A* **70**, 5-6, p. 444.

Lieb, E. H. and Oxford, S. (1981). Improved lower bound on the indirect Coulomb energy, *International Journal of Quantum Chemistry* **19**, 3, p. 427.

Ma, S.-K. and Brueckner, K. A. (1968). Correlation Energy of an Electron Gas with a Slowly Varying High Density, *Physical Review* **165**, 1, p. 18.

Marder, M. P. (2000). *Condensed Matter Physics* (John Wiley & Sons).

Matsukura, F., Ohno, H., Shen, A., and Sugawara, Y. (1998). Transport properties and origin of ferromagnetism in (Ga,Mn)As, *Physical Review B* **57**, 4, p. R2037.

Miyazaki, Y., Igarashi, D., Hayashi, K., Kajitani, T., and Yubuta, K. (2008). Modulated crystal structure of chimney-ladder higher manganese silicides MnSi$_\gamma$ (γ~1.74), *Physical Review B* **78**, 21, p. 214104.

Moore, G. E. (1965). Cramming More Components Onto Integrated Circuits, *Electronics* **86**, 1, p. 114.

Nagaosa, N., Onoda, S., Macdonald, A. H., and Ong, N. (2010). Anomalous Hall effect, *Reviews of Modern Physics* **82**, 2, p. 1539.

Nakayama, H., Ohta, H., and Kulatov, E. (2001). Growth and properties of super-doped Si:Mn for spin-photonics, *Physica B: Condensed Matter* **302-303**, p. 419.

Northrup, J., Hybertsen, M. S., and Louie, S. G. (1989). Quasiparticle excitation spectrum for nearly-free-electron metals, *Physical Review B* **39**, 12, p. 8198.

Ohno, H., Shen, A., Matsukura, F., Oiwa, A., Endo, A., Katsumoto, S., and Iye, Y. (1996). (Ga,Mn)As: A new diluted magnetic semiconductor based on GaAs, *Applied Physics Letters* **69**, 3, p. 363.

Ortiz, G. (1992). Gradient-corrected pseudopotential calculations in semiconductors, *Physical Review B* **45**, 19, p. 11328.

Park, J.-H., Vescovo, E., Kim, H. J., Kwon, C., Ramesh, R., and Venkatesan, T. (1998). Direct evidence for a half-metallic ferromagnet, *Nature* **392**, 6678, p. 794.

Pask, J. E., Yang, L. H., Fong, C. Y., Pickett, W. E., and Dag, S. (2003). Six low-strain zinc-blende half metals: An ab initio investigation, *Physical Review B* **67**, 22, p. 224420.

Pattanaik, A. K. and Sarin, V. K. (2000). Basic Principles of CVD Thermodynamics and Kinematics, in J.-H. Park and T. S. Sudarshan (eds.), *Chemical Vapor Desposition, Surface Engineering Series*, Vol. 2 (ASM International).

Pauling, L. (1938). The Nature of the Interatomic Forces in Metals, *Physical Review* **54**, 11, p. 899.

Perdew, J. P., Burke, K., and Ernzerhof, M. (1996). Generalized Gradient Approximation Made Simple, *Physical Review Letters* **77**, 18, p. 3865.

Perdew, J. P., Jackson, K. A., Pederson, M. R., Singh, D. J., and Fiolhais, C. (1992). Atoms, molecules, solids, and surfaces: Applications of the generalized gradient approximation for exchange and correlation, *Physical Review B* **46**, 11, p. 6671.

Perdew, J. P., Levy, M., and Balduz, J. L. (1982). Density-Functional Theory for Fractional Particle Number: Derivative Discontinuities of the Energy, *Physical Review Letters* **49**, 23, p. 1691.

Perdew, J. P. and Wang, Y. (1992). Accurate and simple analytic representation of the electron-gas correlation energy, *Physical Review B* **45**, 23, p. 13244.

Perdew, J. P. and Zunger, A. (1981). Self-interaction correction to density-functional approximations for many-electron systems, *Physical Review B* **23**, 10, p. 5048.

Pesavento, P. V., Chesterfield, R. J., Newman, C. R., and Frisbie, C. D. (2004). Gated four-probe measurements on pentacene thin-film transistors: Contact resistance as a function of gate voltage and temperature, *Journal of Applied Physics* **96**, 12, p. 7312.

Phillips, J. and Kleinman, L. (1959). New Method for Calculating Wave Functions in Crystals and Molecules, *Physical Review* **116**, 2, p. 287.

Pickett, W. E. and Eschrig, H. (2007). Half metals: from formal theory to real material issues, *Journal of Physics: Condensed Matter* **19**, 31, p. 315203.

Qian, M. C., Fong, C. Y., Liu, K., Pickett, W. E., Pask, J. E., and Yang, L. H. (2006). Half-metallic digital ferromagnetic heterostructure composed of a δ-doped layer of Mn in Si, *Physical Review Letters* **96**, 2, p. 027211.

Qiu, Z. Q. and Bader, S. D. (2000). Surface magneto-optic Kerr effect, *Review of Scientific Instruments* **71**, 3, p. 1243.

Ruderman, M. A. and Kittel, C. (1954). Indirect Exchange Coupling of Nuclear Magnetic Moments by Conduction Electrons, *Physical Review* **96**, 1, p. 99.

Rueß, F. J., El Kazzi, M., Czornomaz, L., Mensch, P., Hopstaken, M., and Fuhrer, A. (2013). Confinement and integration of magnetic impurities in silicon, *Applied Physics Letters* **102**, 8, p. 082101.

Rusz, J., Turek, I., and Diviš, M. (2005). Random-phase approximation for critical temperatures of collinear magnets with multiple sublattices: GdX compounds (X=Mg,Rh,Ni,Pd), *Physical Review B* **71**, 17, p. 174408.

Sandratskii, L. M. (1998). Noncollinear magnetism in itinerant-electron systems: Theory and applications, *Advances in Physics* **47**, 1, p. 91.

Schütz, G., Wagner, W., Wilhelm, W., Kienle, P., Zeller, R., Frahm, R., and Materlik, G. (1987). Absorption of circularly polarized x rays in iron, *Physical Review Letters* **58**, 7, p. 737.

Schwarz, K. (1986). CrO_2 predicted as a half-metallic ferromagnet, *Journal of Physics F: Metal Physics* **16**, 9, p. L211.

Shauhgnessy, M., Damewood, L. J., Fong, C. Y., Yang, L. H., and Felser, C. (2013). Structural variants and the modified Slater-Pauling curve for transition-metal-based half-Heusler alloys, *Journal of Applied Physics* **113**, 4, p. 043709.

Shauhgnessy, M., Fong, C. Y., Snow, R., Liu, K., Pask, J. E., and Yang, L. H. (2009). Origin of large moments in $Mn_x Si_{1-x}$ at small x, *Applied Physics Letters* **95**, 2, p. 022515.

Shauhgnessy, M., Fong, C. Y., Snow, R., Yang, L. H., Chen, X. S., and Jiang, Z. M. (2010). Structural and magnetic properties of single dopants of Mn and Fe for Si-based spintronic materials, *Physical Review B* **82**, 3, p. 035202.

Slater, J. C. (1936). The Ferromagnetism of Nickel, *Physical Review* **49**, 7, p. 537.

Slater, J. C. (1937). Wave Functions in a Periodic Potential, *Physical Review* **51**, 10, p. 846.

Smit, J. (1958). The spontaneous Hall effect in ferromagnetics II, *Physica* **24**, 1–5, p. 39.

Su, W. F., Gong, L., Wang, J. L., Chen, S., Fan, Y. L., and Jiang, Z. M. (2009). Group-IV-diluted magnetic semiconductor $Fe_x Si_{1-x}$ thin films grown by molecular beam epitaxy, *Journal of Crystal Growth* **311**, 7, p. 2139.

Tahir-Kheli, R. A. (1962). Use of Green Functions in the Theory of Ferromagnetism. I. General Discussion of the Spin-S Case, *Physical Review* **127**, 1, p. 88.

Thomas, L. H. (1926). The Motion of the Spinning Electron, *Nature* **117**, 2945, p. 514.

Tinkham, M. (2003). *Group Theory and Quantum Mechanics* (Dover Publications, Mineola, New York).

Tyablikov, S. V. (1959). Retarded and advanced Green functions in the theory of ferromagnetism, *Ukrainskyi Matematychnyi Zhurnal* **11**, p. 287.

Vanderbilt, D. (1990). Soft self-consistent pseudopotentials in a generalized eigenvalue formalism, *Physical Review B* **41**, 11, p. 7892.

Vosko, S. H., Wilk, L., and Nusair, M. (1980). Accurate spin-dependent electron liquid correlation energies for local spin density calculations: a critical analysis, *Canadian Journal of Physics* **58**, 8, p. 1200.

Wang, Y. and Perdew, J. P. (1991). Spin scaling of the electron-gas correlation energy in the high-density limit, *Physical Review B* **43**, 11, p. 8911.

Welkowsky, M. and Braunstein, R. (1972). Interband Transitions and Exciton Effects in Semiconductors, *Physical Review B* **5**, 2, p. 497.

Wiehl, N., Herpers, U., and Weber, E. (1982). Study on the solid solubility of transition metals in high-purity silicon by instrumental neutron activation analysis and anticompton-spectrometry, *Journal of Radioanalytical Chemistry* **72**, 1-2, p. 69.

Wigner, E. and Seitz, F. (1933). On the Constitution of Metallic Sodium, *Physical Review* **43**, 10, p. 804.

Wolf, S. A., Chtchelkanova, A. Y., and Treger, D. M. (2006). Spintronics—A retrospective and perspective, *IBM Journal of Research and Development* **50**, 1, p. 101.

Wu, H., Kratzer, P., and Scheffler, M. (2007). Density-Functional Theory Study of Half-Metallic Heterostructures: Interstitial Mn in Si, *Physical Review Letters* **98**, 11, p. 117202.

Yang, L. H., Shauhgnessy, M., Damewood, L. J., Fong, C. Y., and Liu, K. (2013). Half-metallic hole-doped Mn/Si trilayers, *Journal of Physics D: Applied Physics* **46**, 16, p. 165502.

Yosida, K. (1957). Magnetic Properties of Cu-Mn Alloys, *Physical Review* **106**, 5, p. 893.

Zhang, F. M., Liu, X. C., Gao, J., Wu, X. S., Du, Y. W., Zhu, H., Xiao, J. Q., and Chen, P. (2004). Investigation on the magnetic and electrical properties of crystalline $Mn_{0.05}Si_{0.95}$ films, *Applied Physics Letters* **85**, 5, p. 786.

Zhou, S., Potzger, K., Zhang, G., Mücklich, A., Eichhorn, F., Schell, N., Grötzschel, R., Schmidt, B., Skorupa, W., Helm, M., Fassbender, J., and Geiger, D. (2007). Structural and magnetic properties of Mn-implanted Si, *Physical Review B* **75**, 8, p. 085203.

Zhu, W., Zhang, Z., and Kaxiras, E. (2008). Dopant-assisted concentration enhancement of substitutional Mn in Si and Ge. *Physical Review Letters* **100**, 2, p. 027205.

Zubarev, D. N. (1960). Double-time Green functions in statistical physics, *Soviet Physics Uspekhi* **3**, 3, p. 320.

Index

www.ingramcontent.com/pod-product-compliance
Lightning Source LLC
Chambersburg PA
CBHW050630190326
41458CB00008B/2215